Measuring Livelihoods and Environmental Dependence

Measuring Livelihoods and Environmental Dependence

Methods for Research and Fieldwork

Edited by
Arild Angelsen, Helle Overgaard Larsen, Jens Friis Lund, Carsten Smith-Hall and Sven Wunder

London • Washington, DC

First published in 2011 by Earthscan

Earthscan Ltd, Dunstan House, 14a St Cross Street, London EC1N 8XA, UK
Earthscan LLC, 1616 P Street, NW, Washington, DC 20036, USA
Earthscan publishes in association with the International Institute for Environment and Development

For more information on Earthscan publications, see www.earthscan.co.uk or write to earthinfo@earthscan.co.uk

ISBN: 978-1-84971-132-6 hardback
ISBN: 978-1-84971-133-3 paperback

Typeset by OKS Prepress
Cover design by Clifford Hayes

A catalogue record for this book is available from the British Library

Library of Congress Cataloging-in-Publication Data

Measuring livelihoods and environmental dependence : methods for research and fieldwork / Arild Angelsen ... [et al.].
p. cm.
Includes bibliographical references and index.
ISBN 978-1-84971-132-6 (hb) – ISBN 978-1-84971-133-3 (pb)
1. Household surveys–Developing countries–Methodology. 2. Questionnaires–Developing countries–Methodology. 3. Rural poor–Developing countries. 4. Rural development–Environmental aspects–Developing countries 5. Developing countries–Rural conditions. I. Angelsen, Arild.
HB849.49.M43 2011
001.4'33—dc22 2010047736

At Earthscan we strive to minimize our environmental impacts and carbon footprint through reducing waste, recycling and offsetting our CO_2 emissions, including those created through publication of this book. For more details of our environmental policy, see www.earthscan.co.uk.

Printed and bound in the UK by CPI Antony Rowe.
The paper used is FSC certified.

In memory of Vanessa Annabel Schäffer Sequeira (1970–2006),
a research partner in the Center for International Forestry
Research's Poverty Environment Network
(see Box 9.2 in Chapter 9)

Contents

List of Figures, Tables and Boxes

Figures

Tables

Boxes

Contributors

Arild Angelsen, Professor, Department of Economics and Resource Management, Norwegian University of Life Sciences (UMB), Ås, Norway, and Senior Associate and Poverty Environment Network (PEN) coordinator, Center for International Forestry Research (CIFOR), Bogor, Indonesia. He has done research and published extensively on causes of deforestation, forest and climate (REDD+), and environmental income and poverty. His fieldwork experience is from Indonesia and Eastern Africa. arild.angelsen@umb.no

Ronnie Babigumira, Research Fellow, CIFOR, Bogor, Indonesia. He has done work on land-use change, protected areas and economic growth in Uganda. He has extensive experience working with and teaching statistics using Stata, and is responsible for creating and managing the global data set in PEN. r.babigumira@cgiar.org

Brian Belcher, Professor and Director, Centre for Livelihoods and Ecology, Royal Roads University, Victoria, Canada, and Senior Associate Scientist, CIFOR. His work focuses on understanding and improving the contribution of natural resources to meet development and environmental objectives. He has extensive field experience in Asia and Africa and in Canada. brian.belcher@royalroads.ca

Theresa Bell, Writing Centre Coordinator, Royal Roads University, Victoria, Canada. Her research interests are the methods by which writing centres can help mature students overcome their fears of academic writing, and the use of online writing labs to facilitate writing instruction for distance students. theresa.bell@royalroads.ca

Georgina Cundill, Research Associate, Department of Environmental Science and Post Doctoral Fellow, Department of Environmental Science and the Environmental Learning Research Centre, Rhodes University, Grahamstown, South Africa. She has done research in the areas of social–ecological resilience, collaborative ecosystem management, social learning, multi-scale governance and rural livelihoods. She has fieldwork experience in South Africa, Peru and Chile. georgina.cundill@gmail.com

Amy Duchelle, Research Fellow, CIFOR, Brazil. Her research has focused on community forest management and rural livelihoods in Amazonia with fieldwork experience in Brazil, Bolivia, Peru and Ecuador. She currently coordinates field research for CIFOR's Global Comparative Study on REDD+ in Latin America. a.duchelle@cgiar.org

Sugato Dutt, Doctoral Candidate, Department of Geography, University of Hawaii, Manoa, and Graduate Degree Fellow, East West Center, Honolulu. He has done research on park–people relationships in North Bengal, India. He has also served as a park manager in Tamil Nadu state, India, and a trainer at the Wildlife Institute of India. sugato@hawaii.edu

Pamela Jagger, Assistant Professor, Department of Public Policy and Faculty Fellow, Carolina Population Center, University of North Carolina, Chapel Hill, US, and Senior Associate, CIFOR, Bogor, Indonesia. Her research is focused on the livelihood impacts of natural resource management policies, programmes and projects in sub-Saharan Africa with fieldwork experience in Ethiopia, Uganda and Zimbabwe. pjagger@unc.edu

Helle Overgaard Larsen, Associate Professor, Danish Centre for Forest, Landscape and Planning, University of Copenhagen. She has done research on forest–people interactions, with a focus on livelihood issues, in Nepal and Tanzania. More recently she initiated work on rural livelihoods, community forestry and climate change in Nepal. hol@life.ku.dk

M. K. (Marty) Luckert, Professor, Department of Rural Economy, University of Alberta, Edmonton, Canada, and Research Associate, CIFOR, Bogor, Indonesia. He has done research on livelihoods among poor people using natural resources, with an emphasis on understanding property rights. His fieldwork experience in developing countries is largely from southern Africa. marty.luckert@ualberta.ca

Jens Friis Lund, Associate Professor, Danish Centre for Forest, Landscape and Planning, University of Copenhagen. His research has focused on natural resources decentralization and community forestry, particularly in Tanzania and Nepal. He has recently done research on high forest management and the prospects of REDD+ in Ghana. jens@life.ku.dk

Øystein Juul Nielsen, Post Doc, Danish Centre for Forest, Landscape and Planning, University of Copenhagen. His research concentrates on rural livelihood analyses in developing countries. He is presently doing research on asset dynamics and economic mobility in Nepal. His fieldwork experience also includes eastern Africa. ojn@life.ku.dk

Victoria Reyes-García, ICREA Professor, Institut de Ciència i Tecnologia Ambientals, Universitat Autònoma de Barcelona, Spain. Her research interests include ethnoecology, traditional knowledge systems and community–based conservation. Her fieldwork experience is from Bolivia, Ecuador and India. victoria.reyes@uab.cat

Sheona Shackleton, Senior Lecturer, Department of Environmental Science, Rhodes University, Grahamstown, South Africa. Her research has focused on rural livelihood systems, non-timber forest products and community-based natural resource management, with fieldwork experience from several countries in southern Africa. S.Shackleton@ru.ac.za

Gerald Shively, Professor and University Faculty Scholar, Department of Agricultural Economics, Purdue University, US, and Adjunct Professor, Department of Economics and Resource Management, Norwegian University of Life Sciences (UMB), Ås, Norway. He has conducted survey-based research in more than a dozen countries in Asia, Africa and Latin America on a range of issues related to agricultural development and the environment. shivelyg@purdue.edu

Carsten Smith-Hall, Professor, Danish Centre for Forest, Landscape and Planning, University of Copenhagen and Coordinator for the Forest and Nature for Society global PhD programme and the Sustainable Tropical Forestry joint European MSc programme. His research has focused on commercial utilization of Himalayan biodiversity; current research interests are forest–livelihood relationships, including the role of forests in maintaining human health. cso@life.ku.dk

William D. Sunderlin, Principal Scientist, Governance Programme, CIFOR, Bogor, Indonesia. In CIFOR's Global Comparative Study on REDD, he leads the component on REDD+ project sites. He has conducted research on the underlying causes of deforestation, poverty and well-being, tenure and rights, and climate change. w.sunderlin@cgiar.org

Sven Wunder, Principal Economist and Head of the Brazil office of CIFOR. His main work areas are payments for environmental services, deforestation and forest–poverty linkages. He has published ten books and about 50 academic articles and book chapters. He has advised both small-scale and government conservation programmes, especially in Latin America. s.wunder@cgiar.org

Miriam Wyman, Adjunct Professor, Department of Environment and Society, Utah State University, US. Her research in the Maya Forest region countries of Belize and Mexico has focused on forest livelihoods, the effectiveness of conservation initiatives (for example, nature-based tourism) and protected areas. She is currently involved in a project on best practices for tourism concessions in protected areas worldwide. miriam.wyman@usu.edu

Foreword

Understanding rural livelihoods is one crucial key to putting an end to global poverty. As the authors of this book have demonstrated elsewhere, environmental resources can make up a considerable portion of the livelihood portfolio. But measuring environmental dependence is far from simple, and most of the standard surveys that are undertaken miss many of the environmental resources that are collected, consumed and sold by rural people. With partial surveys comes partial understanding – that will not be the basis of what is needed to drive development and empower rural households.

This book sets out a conceptual framework and method for a deep understanding of rural livelihoods and environmental dependence. It brings together the leading thinkers in this field. It is also the foundation for the global analysis of environmental dependence involving more than 30 PhD students and their supervisors.

This foreword has to end with a personal note. I got to know William Cavendish in the late 1990s when he was doing a highly innovative PhD in rural Zimbabwe. His methods of environmental accounting at household level inspired us to undertake further such studies. Then, in the early 2000s, discussions with colleagues at CIFOR led to the Poverty Environment Network (PEN). I am particularly grateful to Sven Wunder and Arild Angelsen for their considerable work to make PEN a reality, and to the authors for bringing this book to fruition. This book provides a solid methodological foundation for designing and implementing household and village surveys to quantify rural livelihoods, with an emphasis on environmental income and reliance in developing countries.

Bruce Campbell
Director, Program on Climate Change, Agriculture and Food Security (CCAFS) of the Consultative Group on International Agricultural Research

Preface

This book is an output of a large collaborative research project – the Poverty Environment Network (PEN), coordinated by the Center for International Forestry Research (CIFOR) and involving about 30 institutions and 50 individuals across the globe. When PEN was established in 2004, a central aim was to promote better practices for collecting field data to quantify the role of environmental resources in rural livelihoods. But, we quickly realized that the topic goes beyond the allegedly 'simple' issues of questionnaire design and interview techniques, and involves the full research process – from the initial research idea through hypotheses formulation and data collection to analysis and presentation of results. It is not only about asking the questions right, it is also about asking the right questions. This book therefore deals with all the essential steps of the research process.

We would like to sincerely thank all those involved in the PEN project and the preparation of this book. The 33 PEN partners did the hard fieldwork and shared their experiences. These provided valuable lessons that are reflected in the book. Several PEN partners are also chapter or box authors.

The other main group of contributors are PEN resource persons, who have supervised PEN partners, commented on the research tools, attended workshops and responded to various questions arising as PEN went along. Together with the CIFOR scientists involved, they provided the intellectual leadership of the project. Most of the resource persons are chapter authors.

Many chapters have been reviewed by other authors. Finn Helles of the University of Copenhagen gave the penultimate manuscript a critical read and provided valuable suggestions.

The PEN project is co-funded by the Department for International Development (DFID) through the Economic and Social Research Council (ESRC) in the UK, Danida through the Consultative Research Committee for Development Research, and USAID through the AMA-BASIS CRSP programme, and CIFOR core funding. The support from CIFOR management is also acknowledged. PEN started without any secured external funding, based on a belief that 'good ideas eventually get funded'. Several PEN partners also received fieldwork support from the International Foundation for Science (IFS).

Without these various sources of financial support, the PEN project (and this book) would not have materialized.

Finally, we would like to thank Earthscan and Tim Hardwick for their support – and patience – during the publication process.

Arild Angelsen, Helle Overgaard Larsen, Jens Friis Lund, Carsten Smith-Hall and Sven Wunder
Ås (Norway), Copenhagen (Denmark), Rio de Janeiro (Brazil)

October 2010

Acronyms and Abbreviations

BINGO	big international non-governmental organization
CBS	Community Baboon Sanctuary
CI	confidence interval
CIFOR	Center for International Forestry Research
FECONACO	Federación de Comunidades Nativas del Río Corrientes
GPS	global positioning system
GUI	graphical user interface
ICDP	integrated conservation and development project
ICREA	Institució Catalana de Recerca i Estudis Avanfats (Catalan Institution for Research and Advanced Studies)
ICRISAT	International Crops Research Institute for the Semi-Arid Tropics
IFRI	International Forestry Resources and Institutions
IUCN	The World Conservation Union
LSMS	Living Standards Measurement Survey
NGO	non-governmental organization
NTFP	non-timber forest product
OLS	Ordinary Least Squares
OUI	observation unit identifier
PAR	Participatory Action Research
PEN	Poverty Environment Network
PES	payments for environmental services
PLA	Participatory Learning Approach
PRA	Participatory Rural Appraisal
RCT	randomized controlled trial
REDD	Reducing Emissions from Deforestation and Forest Degradation
RRT	Randomized Response Technique
SLF	Sustainable Livelihoods Framework
UAB	Autonomous University of Barcelona
WTA	willingness to accept
WTP	willingness to pay

Chapter 1

Why Measure Rural Livelihoods and Environmental Dependence?

Arild Angelsen, Helle Overgaard Larsen, Jens Friis Lund, Carsten Smith-Hall and Sven Wunder

---

There is in my opinion a right way and we are capable of finding it.
Albert Einstein (1954, *Ideas and Opinions,* Crown Publishers, New York)

The hidden harvest

Measuring rural livelihoods and environmental dependence is not straightforward. Environmental resources are important to millions of poor households in developing countries, yet there is not an established right way to systematically collect data that convey their importance. Such resources, harvested in non-cultivated habitats ranging from natural forests to rangelands and rivers, often contribute significantly to households' current consumption, provide safety nets or pathways out of poverty. The uncertainty regarding the numbers can easily lead to either under- or overestimations (Angelsen and Wunder, 2003). Environmental income often consists of many different and sometimes irregularly collected resources: the forest fruits picked during herding, the medicinal plants collected when grandfather was sick, last month's particularly rich fish catch, and so on. A myriad of resources gathered from multiple sources makes environmental income much harder to recall and quantify than a single annual corn or sorghum harvest. A high share of environmental resources are not traded in markets but consumed directly, further complicating their valuation. The body of literature quantifying environmental resources in rural livelihoods is slowly increasing (for example, Cavendish, 2000; Fisher, 2004; Mamo et al, 2007; Vedeld et al, 2007; Narain et al, 2008; Babulo et al, 2009; Kamanga et al, 2009), but has yet to be widely acknowledged in rural development circles – as becomes evident from recent reviews of rural income and livelihood studies that exclude environmental income (for example, Ellis and Freeman, 2005a).

The general shortage of a representative sample of studies, coupled with the diversity in the quality and methods used in the few existing ones, leave key questions unanswered: how important are environmental resources for poverty alleviation in quantitative terms? When they are important, is it because they can help lift the poor out of poverty or are they mainly useful as gap fillers and safety nets preventing extreme hardship? How do different resource management regimes and policies affect the benefits accruing to the poor? Answers to such questions are essential to design effective policies and projects to alleviate rural poverty. Yet, there is surprisingly little systematic knowledge to answer them adequately.

Published and unpublished quantitative environmental income studies are hard to compare due to methodological differences. In a summary of 54 studies on household environmental income, Vedeld et al (2004, pxiv) noted: 'The studies reviewed displayed a high degree of theoretical and methodological pluralism, and the substantial variability in reporting of specific variables and results is partly explained through such pluralism. This variability must, however, also be attributed to methodological pitfalls and weaknesses observed in many studies.' Methodological challenges include: (a) data generated using long (for example, one-year) recall periods, which is likely to seriously underestimate environmental incomes derived from a myriad of sources (Lund et al, 2008; see also Chapter 7); (b) inconsistent key definitions, for example, what is considered a forest or how income is defined, may differ across studies, making findings incomparable; (c) a host of survey implementation problems, such as failure to adequately train enumerators or check data while in the field, resulting in questionable data quality; and (d) a widespread perception that it is too difficult and costly to obtain high quality environmental income data. The geographical coverage of available studies is also limited, with most coming from dry southern and eastern Africa. Thus, while our knowledge regarding environmental income and rural livelihoods is incrementally improving, we believe that more in-depth studies across a range of sites are required, preferably using best-practice and unified methodologies that enable comparison and synthesis. This book is designed to be an instrument to help make it happen.

Designing and implementing household and village surveys for quantitative assessment of rural livelihoods in developing countries is challenging, with accurate quantification of income from biologically diverse ecosystems, such as forests, bush, grasslands and rivers, being particularly hard to achieve. However, as the above published studies indicate, this 'hidden harvest' (Scoones et al, 1992; Campbell and Luckert, 2002) is too important to ignore. Fieldwork using state-of-the-art methods and, in particular, well-designed household questionnaires, thus becomes an imperative to adequately capture environmental income dimensions of rural welfare. In fact, current poverty alleviation strategies in most developing countries draw to a significant extent on results from household surveys; environmental income estimates are, however, often not included in the standardized living standards measurement surveys (Oksanen and Mersmann, 2003). Studies based on such surveys are thus inadequate for understanding the diversity of rural income generation in developing countries.

One attempt to overcome this shortage of data is a large global-comparative research project: the Poverty Environment Network (PEN), described later in this chapter. The book draws widely on the methodological experiences from PEN.

Purpose of this book

This book aims to provide a solid methodological foundation for designing and implementing household and village surveys to quantify rural livelihoods, with an emphasis on quantifying environmental income and reliance in developing countries. All the major steps are covered, from pre-fieldwork planning, in-the-field sampling, questionnaire design and implementation, to post-fieldwork data analysis and result presentation. The intention is to provide input to the entire research process, in the specific context of developing and implementing operational research ideas using quantitative approaches in developing countries, while limiting the 'remember the malaria pills' and 'get to know the local culture' generalized advice that is covered well in other books.

The book is aimed at: (a) graduate students and researchers doing quantitative surveys on rural livelihoods, including (but not limited to) the environment, and (b) practitioners in government agencies, international aid agencies and non-governmental organizations (NGOs) involved in fieldwork in relation to project implementation (for example, baseline and impact studies) in developing countries.

The book relates to three broad groups of literature: (a) the limited body of works at the graduate student level on broad field methods and introductions to the practicalities of fieldwork (for example, Barrett and Cason, 1997; Scheyvens and Storey, 2003); (b) the extensive literature on quantitative surveys in general, including work on business research methods (for example, Bryman and Bell, 2003) and measurement of living standards (for example, Grosh and Glewwe, 2000); and (c) the emerging body of literature on livelihoods in developing countries (for example, Ellis, 2000; Ellis and Freeman, 2005b; Homewood, 2005).

This book has four major distinctive features in relation to this pre-existing body of knowledge. First, it fills a gap between the three types of literature by giving a thorough review of using quantitative methods in rural livelihoods studies in developing countries. Second, it deals not only with quantitative household and village surveys, but covers the entire research process, as opposed to books focusing on methods (for example, Foddy, 1993), analysis of survey data (Deaton, 1997), fieldwork practicalities (for example, Barrett and Cason, 1997) or livelihood case studies (for example, Homewood, 2005). Third, it centres on rural livelihoods with environmental dependence as the predominant example of how to deal with livelihoods complexity, as opposed to other approaches that focus on issues such as health, education and agricultural income while disregarding information on environmental incomes (for example, Grosh and Glewwe, 2000). Fourth, as explained in the next section, it draws on the extensive comparable experiences in the PEN project of more than 50 researchers who have implemented and supported quantitative household and village surveys across a variety of continents, countries and cultures.

The Poverty Environment Network (PEN)

PEN is an international network and research project on poverty, environment and forest resources, organized by the Center for International Forestry Research (CIFOR). It was established in 2004 in order to address the environmental income issues identified in the introduction to this chapter, which have particular relevance for natural forests. The core of PEN is a tropics-wide collection of uniform socio-economic and environmental data at the household and village levels, undertaken by 33 PEN partners (mainly PhD students) and supported by some 20 PEN resource persons (CIFOR researchers, associates and external university partners acting as supervisors with active field presence), jointly generating a global database with more than 9000 households from 25 countries. The PEN project is arguably the most comprehensive study done in the field of poverty and environment, and will serve as the basis for the first global-comparative and quantitative analysis of the role of tropical environmental resources in poverty alleviation. This book is written by PEN partners and resource persons who have been involved in the design and implementation of PEN, as well as dozens of other projects collecting similar data.

PEN is built on the observation that some of the best empirical data collection is done by PhD students: they often spend long periods in the field and personally supervise the data collection process, thereby getting high quality data – something that established university researchers often lack the time to achieve. The basic PEN idea was to achieve two goals: (a) use the same ruler (same questionnaires and methods) to make data comparable, and (b) promote good practices and thus increase the quality of data. In short, the value added of the individual studies can be substantially enhanced by using standardized and rigorous definitions and methods, which permit comparative analysis.

The first phase of PEN from 2004 to 2005 focused on identifying and designing the research approach and the data collection instruments and guidelines, and at the same time building up the network through PEN partner recruitment. Fieldwork and data collection by PEN partners started in 2005. After completion in 2009, a third phase of data cleaning, establishing the global data set, data analysis and writing began. The project is expected to be completed in 2012, and the PEN data set will eventually be made publicly available for use by researchers.

The PEN research approach

During early discussions and workshops, a consensus quickly emerged that, in order to get reliable estimates of environmental resource uses, a detailed recording (income accounting) method was needed, using short recall periods – one year being far too long for the accuracy aimed for. This was particularly inspired by work done in Zimbabwe by Cavendish (2000) and Campbell et al (2002). It was also decided that PEN data collection should consist of three types of quantitative surveys (in addition to an attrition and temporary absence survey) covering a full year:

- Two village surveys (V1, V2).
- Two annual household surveys (A1, A2).
- Four quarterly household surveys (Q1, Q2, Q3, Q4).

The timing of the surveys is shown in Figure 1.1. Data collection requires a fieldwork period of no less than ten months. The village surveys (V1–V2) collect data that are common to all households, or show little variation among them (cf. Chapter 6); V1 is done at the beginning of the fieldwork to get background information on the villages, while V2 is done at the end of the fieldwork period to get information for the 12 months of accumulated recall period covered by the surveys. The household surveys were grouped into two categories: (a) annual household surveys, with A1 at the beginning of fieldwork providing household information serving as a baseline (demographics, assets, forest-related information), while A2 at the end collected information for the 12 month period covered by the surveys (for example, on risk); and (b) the four quarterly

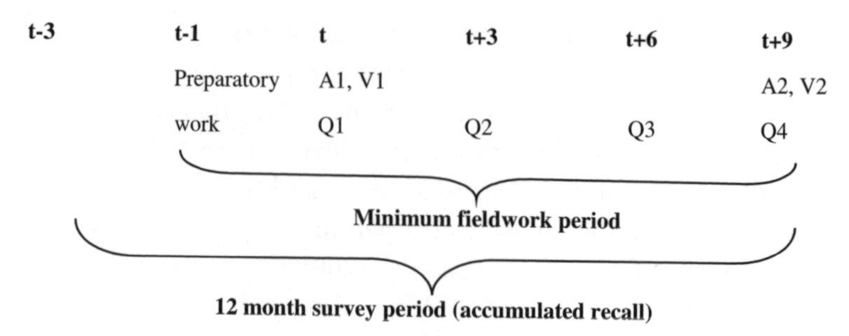

Figure 1.1 *The timing of village and household surveys in a PEN study*

household surveys that focused on collecting detailed income information. All research tools (the prototype questionnaires and the associated technical guidelines; the template for data entry; the codebook; and the data cleaning procedures) can be downloaded from the PEN website (www.cifor.cgiar.org/pen). Prototype questionnaires are available in English, French, Spanish, Portuguese, Chinese, Indonesian, Nepalese and Khmer. While PEN pursued a common methodology, all prototype questionnaires were pretested and adapted to local conditions at each research site. Each PEN partner submits his/her final data set, along with a narrative adhering to a standard template and providing detailed contextual site information, to the global database.

A key feature of the PEN research project is the collection of high quality data through the quarterly household surveys. These include detailed data collection on all types of income, not just environmental sources. In addition to the higher accuracy and reliability of quarterly income surveys, various income-generating activities often have considerable seasonal variations, and documenting these can help us understanding fluctuations and seasonal gap fillers. The recall period in the quarterly income surveys was generally one month, which would then be extrapolated to the three-month period. The exception was agricultural income and 'other income' (remittances, pension, and so on) that used three months, as these are major income sources (easier to remember) and might be irregular (thus the full 12-month period is covered). The PEN technical guidelines also emphasize that all major products with irregular harvesting, for example, short-lived mushrooms harvested for sale on a large scale or the occasional sale of a timber tree from private land, should be identified early on, for example, during preparatory fieldwork and pretesting of questionnaires. A one-month recall in quarterly surveys entails the risk of missing out on these activities, thus a three-month recall was applied for such products. In general, the recall period has to be selected optimizing a trade-off between completeness and accuracy in the respondents' recall, which will vary across economic activities (Chapter 7).

Coverage and selection of PEN study sites

PEN sites cover major sub-continental areas in Africa, Asia and Latin America (Figure 1.2). Being a collaborative research project, PEN did not have the authority, nor the resources, to fully determine the location of the individual PEN study sites or villages. However, in general study localities were chosen so as to: (a) display at least a minimum level of forest dependence; (b) meet criteria relevant to the topics of each individual study (going beyond the core of PEN); and (c) meet PEN's site sampling criteria of representation and variation (Chapter 4). Globally, we had study site gaps in West Africa and Indochina that were subsequently filled in through targeted external fund-raising and collaboration with partners in these regions. Regarding representation, the aim was to avoid too many special cases, for example, areas with unusually valuable forest products, unusually favourable or unfavourable conditions for income generation, or a history of very heavy donor intervention. The lack of centrally planned study site selection implies, of course, that one *cannot* draw generalized conclusions for a country from one single PEN study site, for example, stating that the forest income share in Zambia is identical to the one found in our PEN case study.

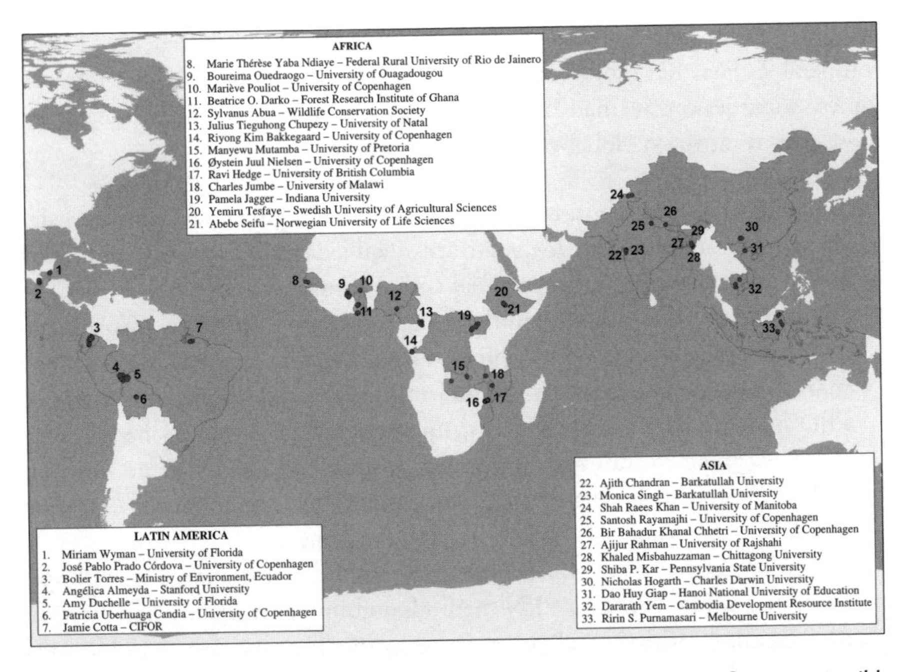

Figure 1.2 *Geographical location of the PEN study sites and lists of site-responsible scientists*

To achieve a representative sample, PEN paid particular attention to variation within study areas. We searched explicitly for variation along key gradients (such as market distance, vegetation types, land tenure and institutions, population density and growth, predominance of ethnic groups and commercial stakeholders, sources of risk and levels of poverty) to make site results representative of a larger universe – for example, a district or province of the country being studied (Cavendish, 2003). Not all these gradient variations are found within any single study area and gradients often correlate, making choices easier: market-remote areas tend to be poorer yet richer in natural vegetation, less densely populated yet with a higher share of indigenous people. PEN partners were advised by their PEN supervisors during start-up workshops on how to choose study areas, and selection of villages within those areas.

To be included in the global PEN data set, a minimum sample of 100 households was required. Most studies had higher samples, with an average of about 240–250 households. While PEN sites and villages were thus selected according to explicit stratification criteria, the within-village selection of households followed random sampling, using household lists and pre-existing censuses (see Chapter 4).

Lessons learned from PEN planning and implementation

The basic PEN idea, to develop a common set of methods and establish a network of primarily PhD students to generate a critical pool of high quality and comparable data, has scope for being replicated in relation to other research topics. Some lessons learned from PEN may therefore have wider interest for those who want to do global-comparative research:

- A high-profile research institution should be leading such an effort in order to sell and market the idea, attract qualified PhD students and their supervisors, as well as other external collaborators.
- Allow for sufficient time to jointly develop common prototype instruments and technical guidelines, developing and agreeing on these may take a year or more. This key initial activity should be well advanced before recruitment of PhD students, who otherwise may drop out because their time schedule does not allow them to wait for the research instruments. At the same time, an inclusive process in developing and modifying methods will increase partners' ownership and commitment to the project.
- Establish meeting places for project partners. PEN had annual workshops, the initial ones focused on PEN methods and implementation; and later on data cleaning, analysis and presentation of preliminary results. Additional fora included an electronic news letter and a mailing list. Web-based discussion groups could also be used (not done in PEN).

- A global–comparative project needs to provide tangible benefits to the partners (in exchange for the data provided): networking, sponsored participation in workshops and conferences, joint discussion of methods, standardized quality check and feedback on collected data, supervision from resource persons, assistance in funding applications and advice on data analysis and writing. The drawbacks also need to be communicated, for example, strict centralized quality control and pressure to deliver data on time.
- Allow ample time for data checking, cleaning, and harmonization of comparative standards; this takes more time than planned in 99 per cent of cases. Standardized central data quality control procedures should be funded and established; individual timelines should be established for data submission.

PEN also *ex post* conducted a survey among PEN partners to evaluate the PEN prototype questionnaires and lessons learned from the field. Five areas stand out in the responses:

- Stay in the field as much as possible. The field presence is essential to build trust, collect contextual information and supervise the data collection. This is indeed one reason why field data by PhD students are among the best – more senior researchers often cannot set aside sufficient time to be present in the field.
- Have a dedicated team of enumerators. The enumerators and field assistants are critical for successful fieldwork. Identify and select good enumerators, train them, pay them reasonable salaries, be clear in communication and boost morale by regular interactions and team-building exercises. Being in the field and arranging social activities were the two most common ways to maintain enumerator motivation. However, random checks of performance may also be necessary to discourage sloppiness and data falsification.
- Building relations with respondents is the key to success. Explain the purpose of the research to the local people, respect them and get to know them – 'Hang out with people, celebrate with them, play with their children.' Spend as much time as possible talking to the local people 'since the qualitative understanding is critical to understanding/interpreting quantitative results'.
- Check and double-check data. Check questionnaires as soon as possible after the interviews, correct errors, enter data early (preferably in the field). Also, do not underestimate the time required for data cleaning and management post-fieldwork, it can be grossly underestimated.
- Have a plan – and a plan B. 'Flexibility, patience, courage, determination and a sense of adventure is essential to successfully pull off something like this.' Field research is a logistical challenge.

Structure and content of this book

This book focuses on the design and implementation of surveys – as opposed to other research methods, such as experiments, archival analysis or case studies. Surveys gather information by asking questions of respondents. This method has advantages when focusing on contemporary events, when no control of behavioural events is required, and when answering categorical research questions of the what/who/where/how many/how much type (Yin, 2009). In particular, this book is focused on questionnaire surveys that are administered by enumerators through personal interviews with respondents. The book covers the entire research process, from generating research ideas and hypotheses to collecting and analysing data, and communicating the results. Figure 1.3 presents a schematic overview of the relationship between the research process and chapter content. Chapters 2–8 focus on activities and considerations that should start before fieldwork, Chapters 9–11 on activities and issues during fieldwork, and Chapters 12–14 on post-fieldwork activities. In practice, the research process is iterative and facts disclosed during the preparatory field stages may require revisions in research design. Hence, interpretation of the stylized figure should allow for various feedback flows.

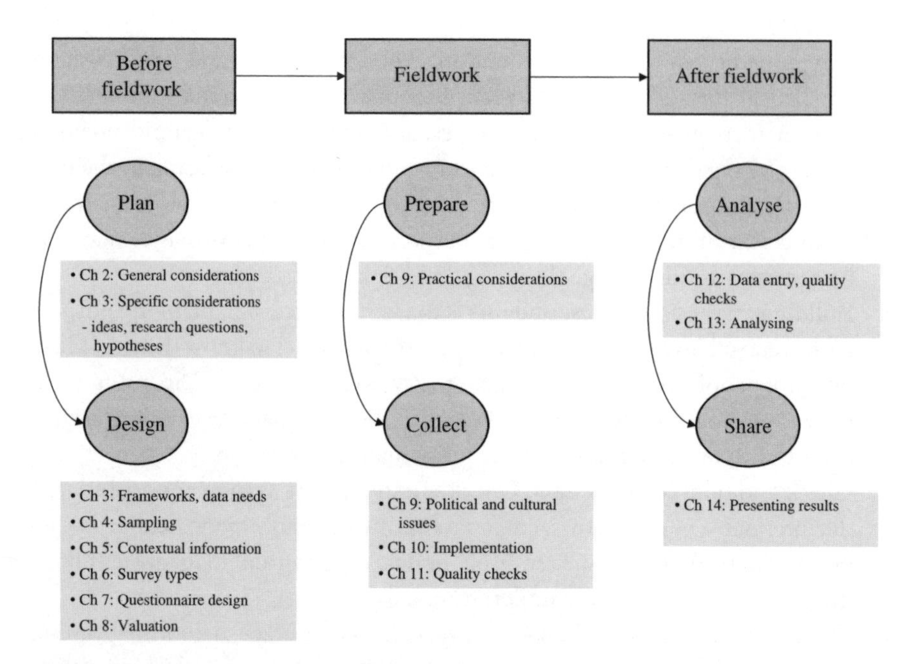

Figure 1.3 *Overview of the relationship between the research process and chapter content*

The type of field research described in this book involves collecting large amounts of empirical data in developing countries and can be demanding in terms of finances, researcher time and respondent patience. Therefore, before setting out, it is worthwhile to consider the justifications for going. In Chapter 2, Victoria Reyes-García and William D. Sunderlin argue that field research is justifiable because it can increase the scientific understanding of a problem through providing access to data not otherwise sufficiently available, deeper contextual understanding, enhanced data quality and inspiration to challenge conventional wisdom. Field research also enables identification of locally perceived problems, as well as insertion of local views into the policy process. The chapter further provides an overview of the various interests involved in evaluating the merits of research. It encourages researchers to be mindful of local research priorities, to involve local people in the research conceptualization and design, and to reflect on sources of personal motivation.

Having decided to do field-based research, the research project starts with the development of a proposal. According to Arild Angelsen and co-authors of Chapter 3, the research proposal should answer two essential questions: *what* will be investigated, and *how*. The chapter outlines eight essential steps and describes the process of developing research ideas, objectives, questions and hypotheses that are based on the theoretical foundation and empirical evidence of a scientific field. Furthermore, the proposal needs to identify the data needed to answer the research questions and test the hypotheses, and state how the data are to be generated and analysed. The chapter offers a schematic framework for developing a logically coherent research design, emphasizing the need for consistency between the 'what' and 'how' questions.

Deciding on who and how many to include in a sample – and how to select them – is fundamental to any empirical research. In Chapter 4, Gerald Shively argues that the decision regarding a sampling strategy must take the research questions and hypotheses as its point of departure, as well as an identification of the target population for the research, in other words, what population the results should be representative of. The sampling strategy should allow both for variation that enables one to answer the research questions and for generalization to the target population. The chapter provides examples of probability and non-probability sampling procedures, as well as rules of thumb and more formal procedures to decide on a sample size.

Continuing with the design of the research, contextual information will both inform and complement quantitative household surveys. In Chapter 5, Georgina Cundill and co-authors use the Sustainable Livelihoods Framework to discuss types of contextual information useful for situating and understanding livelihood strategies and for preparing a household survey. Types of data and potential collection methods are presented for five livelihood assets; policy, institutions and processes; and the vulnerability context. The authors emphasize the need for the researcher to spend time discussing with and listening to people, trying to understand their perceptions concerning what is important.

Having defined what data are needed, the researcher will continue to select methods for data collection. In order to collect data at the appropriate level (for example, not overload household questionnaires) and format, Pamela Jagger and Arild Angelsen in Chapter 6 present a framework for deciding which surveys to undertake. The choices of scale (household versus village) and survey format depend on two key questions: (a) at which level does the variable vary, and (b) are representative quantitative figures needed in the later analysis? To complement other chapters focusing on the household level, the chapter provides approaches for detailed description of village-level data collection and outlines different village-level sources of data. Other types of surveys covered are surveys of prices and wages, value chains, local institutions and groups.

When it is clear what data needs should be covered by the household survey, the next step is to design the questionnaire and formulate specific questions. Overarching Chapter 7, by Arild Angelsen and Jens Friis Lund, is the need to decompose and translate overall data needs into questions that can be meaningfully answered by respondents without compromising construct validity in the process. At the end of this process, one should have a questionnaire consisting of a set of logically ordered questions that are concrete, specific, simple and using neutral formulations and carefully defined terms. Accordingly, the chapter provides guidelines on this decomposition and translation process, as well as on design and formulations of the specific questions.

Accurate and reliable estimation of non-marketed environmental product values is essential in order to generate trustworthy income data. In Chapter 8, Sven Wunder and co-authors describe some of the structural obstacles found in rural economies of developing countries and then proceed to outline and review six different practical methods for how to assign values to non-marketed goods. They end the chapter by introducing techniques for checking the validity and reliability of values.

With a good research design in hand, it is time to embark on the field data collection. In Chapter 9, Pamela Jagger and co-authors use their PEN fieldwork experience to provide suggestions for how to make fieldwork a fruitful and pleasant experience by spending time attempting to understand the political context, understanding and behaving respectfully within the cultural context and planning the practicalities concerning the research team's stay in the study area. The chapter argues that the study of natural resources is fraught with political complexity, emphasizes the need to conduct the research transparently and share findings with the local communities, and discusses health and safety aspects of staying in the study area.

The considerable time and efforts spent on preparation should ensure that the conditions for fieldwork are optimal. To then make the actual data collection a success – in other words, to collect data of high quality – one factor of utmost importance is the quality of the enumerators hired to implement the survey. In Chapter 10, Pamela Jagger and co-authors address the issues of selection, training and upholding the continued motivation of enumerators, aspects of field research often omitted from survey method descriptions. A crucial part of survey fieldwork is the management of the questionnaires used to capture the data. The chapter suggests ways of storing and keeping track of questionnaires, and provides a protocol for checking the questionnaires in the field.

Getting quality data is a major focus of this book. In Chapter 11, Jens Friis Lund and co-authors focus on how data quality may be affected by systematic measurement errors, in other words, errors arising during the implementation of the survey that systematically affect the measurement of a variable across a sample. The multiple reasons for systematic measurement errors are outlined along with procedures on how to avoid or minimize them, focusing on issues related to enumerators and questionnaire administration, respondents' strategic behaviour and understanding, and bounded knowledge.

In Chapter 12, Ronnie Babigumira addresses an aspect as important to data quality as sample selection, questionnaire design and enumerator selection and training: data entry and checking. The chapter cautions against combining the various steps of data entry and analysis and provides a thorough discussion of how best to code, enter and check data. Examples of typical errors illustrate the importance of the chapter's advice, and the advantage of using codes rather than storing data as text is emphasized.

After the collecting, entering, checking and cleaning of the data, the researcher should be, finally, ready to start the analyses that may answer the research questions initially set out. Gerald Shively and Marty Luckert in Chapter 13 suggest approaches to data analysis, starting with exploration of data using descriptive statistics, and moving on to hypothesis-driven analyses using multiple regression analysis to establish cause-and-effect relationships. The importance of formulating unambiguous hypotheses, based on theory or empirical studies, with accurately defined variables is emphasized, and common pitfalls in data interpretation are presented.

The last steps of the research process are the write-up and communication of the findings with the purpose of 'making a difference', in other words, to influence policies and practices. In Chapter 14, Brian Belcher and co-authors argue that research can influence policies through multiple and iterative pathways, including identification and involvement of the intended audience already in the research design phase, and the conscious use of other media complementing the scientific paper. Recognized as the standard format for communication of research results, the structure of the scientific paper is outlined and suggestions for efficient writing are made.

References

Angelsen, A. and Wunder, S. (2003) 'Exploring the forest-poverty link: Key concepts, issues and research implications', CIFOR Occasional Paper no 40

Babulo, B., Muys, B., Nega, F., Tollens, E., Nyssen, J., Deckers, J. and Mathijs, E. (2009) 'The economic contribution of forest resource use to rural livelihoods in Tigray, Northern Ethiopia', *Forest Policy and Economics*, vol 11, pp109–117

Barrett, C. B. and Cason, J. W. (1997) *Overseas Research*, Johns Hopkins University Press, London

Bryman, A. and Bell, E. (2003) *Business Research Methods*, Oxford University Press, Oxford

Campbell, B. M. and Luckert, M. K. (eds) (2002) *Uncovering the Hidden Harvest: Valuation Methods for Woodland and Forest Resources, People and Plants*, Earthscan, London

Campbell, B. M., Jeffrey, S., Kozanayi, W., Luckert, M., Mutamba, M. and Zindi, C. (2002) *Household Livelihoods in Semi-arid Regions: Options and Constraints*, CIFOR, Bogor

Cavendish, W. (2000) 'Empirical regularities in the poverty-environment relationship of rural households: Evidence from Zimbabwe', *World Development*, vol 28, no 11, pp1979–2000

Cavendish, W. (2003) 'How do forests support, insure and improve the livelihoods of the rural poor? A Research Note', unpublished paper, CIFOR, available at www.cifor.cgiar.org/pen

Deaton, A. (1997) *The Analysis of Household Surveys: A Microeconometric Approach to Development Policy*, Johns Hopkins University Press, Baltimore, MD

Ellis, F. (2000) *Rural Livelihoods and Diversity in Developing Countries*, Oxford University Press, Oxford

Ellis, F. and Freeman, H. A. (2005a) 'Comparative evidence from four African countries', in Ellis, F. and Freeman, H. A. (eds) *Rural Livelihoods and Poverty Reduction Policies*, Routledge, London, pp31–47

Ellis, F. and Freeman, H. A. (eds) (2005b) *Rural Livelihoods and Poverty Reduction Policies*, Routledge, London

Fisher, M. (2004) 'Household welfare and forest dependence in Southern Malawi', *Environment and Development Economics*, vol 9, pp135–154

Foddy, W. (1993) *Constructing Questions for Interviews and Questionnaires: Theory and Practice in Social Research*, Cambridge University Press, Cambridge

Grosh, M. and Glewwe, P. (2000) *Designing Household Survey Questionnaires for Developing Countries: Lessons from 15 Years of the Living Standards Measurement Study*, vols 1–3, World Bank, Washington, DC, http://go.worldbank.org/NTQLJEEXQ0, last accessed 5 February 2011

Homewood, K. (ed) (2005) *Rural Resources and Local Livelihoods in Africa*, James Currey Ltd, Oxford

Kamanga, P., Vedeld, P. and Sjaastad, E. (2009) 'Forest incomes and rural livelihoods in Chiradzulu District, Malawi', *Ecological Economics*, vol 68, pp613–624

Lund, J. F., Larsen, H. O., Chhetri, B. B. K., Rayamajhi, S., Nielsen, Ø. J., Olsen, C. S., Uberhuaga, P., Puri, L. and Prado, J. P. P. (2008) 'When theory meets reality: How to do forest income surveys in practice'. Forest & Landscape Working Paper No 29-2008, Centre for Forest, Landscape and Planning, University of Copenhagen, Copenhagen

Mamo, G., Sjaastad, E. and Vedeld, P. (2007) 'Economic dependence on forest resources: A case from Dendi District, Ethiopia', *Forest Policy and Economics*, vol 9, pp916–927

Narain, U., Gupta, S. and Veld, K. (2008) 'Poverty and resource dependence in rural India', *Ecological Economics*, vol 66, pp161–176

Oksanen, T. and Mersmann, C. (2003) 'Forests in poverty reduction strategies: An assessment of PRSP processes in sub-Saharan Africa', in Oksanen, T., Pajari, B. and Tuomasjukka, T. (eds) *Forests in Poverty Reduction Strategies: Capturing the Potential*, EFI Proceedings No 47, pp121–155

Scheyvens, R. and Storey, D. (2003) *Development Fieldwork: A Practical Guide*, Sage, London

Scoones, I., Melnyk, M. and Pretty, J. N. (1992) *The Hidden Harvest: Wild Foods and Agricultural Systems: A Literature Review and Annotated Bibliography*, International Institute for Environment and Development, London

Vedeld, P., Angelsen, A., Sjaastad, E. and Berg, G. K. (2004) 'Counting on the environment: Forest income and the rural poor', *Environmental Economics Series 98*, World Bank, Washington, DC

Vedeld, P., Angelsen, A., Bojö, J., Sjaastad, E. and Berg, G. K. (2007) 'Forest environmental incomes and the rural poor', *Forest Policy and Economics*, vol 9, pp869–879

Yin, R. K. (2009) *Case Study Research: Design and Methods*, Applied Social Research Methods Series vol 5, Sage, London

Chapter 2

Why Do Field Research?

Victoria Reyes-García and William D. Sunderlin

The aim of science is not to open the door to infinite wisdom, but to set a limit to infinite error.
Bertolt Brecht, *The Life of Galileo* (1939, scene 9)

Introduction

Field research is a methodological approach to observe behaviour under natural conditions. Field research is traditionally contrasted to research conducted in laboratories or academic settings, or to research exclusively relying on existing, or secondary, data. In the social sciences, the collection of raw data *in situ* often – but not exclusively – occurs in a geographical and cultural context not familiar to the person collecting the data. Differently from other methodological approaches, field research in the social sciences allows the researcher to engage in detailed observation and conversations that give the opportunity to elicit information regarding the data being collected. Many techniques and methods for data collection can be used during field research (Bernard, 1995), including:

- **Observation** of events as they occur in natural settings sometimes expanded by means of a contextual inquiry. Observation can be naturalistic or participant, when the researcher engages in the observed activities.
- **Archival research** or the study of information from already existing records, such as national census or local publications, but also personal documents.
- **Field experiments** or experiments conducted in natural settings in order to understand causal relations among phenomena.
- **Surveys** or the collection of systematic data on people's actions, thoughts and behaviour through asking direct questions in natural settings.

In the next section, we outline reasons that justify the investment in field research in general. Then, in the section that follows we ask why one should do field research in poor developing countries. We pose the question at three levels: the interest of society; the interest of the community being researched; and the interest of the researcher. Before concluding, we also discuss some of the ethical challenges related to doing field research. This last section helps one understand how to prepare and carry out field research properly, if one should decide to do it.

Four basic reasons for doing field research

Field research has been a common technique in the social sciences during most of the 20th century (see Box 2.1). But field research, including the collection of data through household surveys – the main method discussed through this book – can be expensive, time-consuming and, in some cases, invasive. Who likes to have strangers ask personal questions concerning your level of education, the number of chickens on your farm, possibly illegal uses of the forest and the amount of remittance income you got from your daughter who lives abroad? So what is it about field research that justifies the often extraordinary amount of effort involved in conducting it, and especially in doing it well? After all, tons of data – including household-level data from developing countries – can be downloaded in a few minutes and free of charge from the internet. Why, despite the high costs in time and money, have researchers from many disciplines adopted field research as a valid methodological approach for collecting data? We outline four basic reasons:

Overcoming lack of data

Field research is often necessary to fill an information void related to the problem to be investigated. Often there is very little or no existing information concerning a problem in a given place or given topic. The problem might be known or suspected by hearsay and rumour, or through reports in newspapers and on the radio, but without primary data to analyse it in a scientific and systematic way. If there is information concerning the problem in the national census, it might be inadequate for gaining insights on its cause, development and possible resolution. For example, national census information is often available at high levels of aggregation only. Field research allows us to test theories at a low level of aggregation because field researchers typically collect information on some of the basic units for decision-making parameters (communities, household, persons). Even when some amount of data exists, gaps might need to be filled. In that case, one could conduct targeted supplementary field research to collect the complementary data needed. Field research makes possible the scientific exploration of problems in geographic

Box 2.1 *The birth of fieldwork*

Anthropologists attribute the development of the modern tradition of field research to Malinowski, through his study of the Trobriand Islanders of New Guinea (Malinowski, 1922). Malinowski argued that anthropologists needed to get off the verandas of the missionaries' and government officials' houses to see what local people were really doing. The basic idea was that, only by immersing oneself in people's daily activities and talking to local people in their homes and fields, could one hope to understand them.

> Indeed, in my first piece of ethnographic research on the South coast, it was not until I was alone in the district that I began to make some headway; and, at any rate, I found out where lay the secret of effective field-work. What is then this ethnographer's magic, by which he is able to evoke the real spirit of the natives, the true picture of tribal life?...
>
> Field-work consists mainly in cutting oneself off from the company of other white men, and remaining in as close contact with the natives as possible, which really can only be achieved by camping right in their villages... And by means of this natural intercourse, you learn to know him, and you become familiar with his customs and beliefs far better than when he is a paid, and often bored, informant. (Malinowski, 1922, pp6–7).

For many years, field research was the most common – and sometimes the only – methodological approach of cultural anthropologists. Once mainly a domain of anthropologists, field research is now widely conducted in most of the social sciences, including geography, sociology and economics (Udry, 2003).

areas or on research topics where there are few pre-existing data. Indeed, the major rationale for the PEN project was the lack of comparable data on the relationship between forests and poverty (Chapter 1).

Understanding the context

Even in cases where there is a perfect set of available data to answer a research question, researchers opt to conduct complementary field research. Economists, for example, often conduct short field research visits to understand the social and economic context of the location where the data were collected (see, for example, the work of Pender (1996) in the International Crops Research Institute for the Semi-Arid Tropics (ICRISAT) Village Level Studies). For example, imagine that you use information from the national census to study a region's economy. You find that most people derive their livelihoods from agriculture and that most land is communal. But you also find a high inequality of income in agriculture. The finding is puzzling: why is there so much inequality if land is held in common? Field research can help you understand the context of your findings. It might be possible that, because there are high taxes on private land ownership but no taxes of communal land ownership, people declare lands as communal (to avoid taxes) but use them privately in accordance with customary rules for land distribution.

Field research can thus provide a deeper understanding of the local situation, allowing the researcher to measure the origins, scope and scale of a problem, as well as to gauge local opinions on the causes, consequences and means to resolve a problem. In the best of cases, with a large and representative sample of households, it might be possible for research results to serve as an input for rethinking or guiding policy at the national level. But even short of this, local case study research might provide vital insights for understanding and resolving a pressing problem.

Controlling data quality

Field research enables control of the accuracy of data collection through at least two mechanisms. First, field research enables corroboration or confirmation of data via triangulation (see Chapter 11). For example, answers to household surveys can be checked against information from other interviewees, observation or written records locally available. Field research helps the researcher determine which results are valid. Second, field research enables the researcher to select sensible questions for the specific cultural context being investigated (for example, to avoid asking Muslim respondents about pork meat consumption).

Furthermore, cultural or ethnic differences can affect the interpretation of a question, but people's willingness to give accurate answers might also vary depending on their trust of the interviewer. For example, in a culture where government and/or business are perceived as being corrupt or exploitive, responses to questions from outsiders are likely to be affected by the local perception that responses may be obtained and abused by government officials or others. As discussed later in this book (Chapter 11), fieldwork can improve the quality of the data collected by: (a) increasing the trust of people in the researcher, and (b) allowing questions to be identified that might be sensible in a given cultural context, as well as improving the way those questions are being asked.

Opening new frontiers of knowledge

Observing the local reality often tells you things that cannot be observed through national census or survey data. Field research thus puts researchers in contact with a situation that can open their eyes and enable them to initiate new lines of thinking. Field research can provide an empirical basis (and, in some cases, the only basis) for challenging conventional wisdom or for testing a research question, a theoretical proposition or a hypothesis related to a pressing issue.

Reasons for doing field research in poor developing countries

Why should we do field research in poor developing countries? To answer this question it is appropriate to frame the issue in a larger context: Why should one do *research* in poor countries, not just field research but also the entire linked research enterprise – including research conceptualization and design, bibliographical research, analysis of the census and other national data, and the like? A preliminary answer to these questions is rather obvious. Poor developing countries are places that are often beset by many problems including:

- Low income, livelihood insecurity, vulnerability and poverty.
- Insufficient and unreliable access to health care and education.
- Lack of voice and power of ordinary people in the national and local policy.
- Gender oppression and inequality.
- Lack of access to markets with subsequent low prices for the produce.
- Inadequacy and unfairness of laws and regulations.
- Victimization of local people by powerful outside entities (for example, government, military, private enterprises).
- Lack of recognition of rights, including: tenure over land and resources (customary and/or statutory); citizenship; civil rights; human rights.
- Problems related to environmental management and conservation (for example, deforestation, restriction to access natural resources, climate change).
- Conflict and war.
- Natural disasters (for example, earthquakes, droughts, hurricanes, tsunamis) and epidemics (for example, HIV/Aids, ebola virus).

It is important to point out that *all* of these problems exist to a degree in so-called developed countries, making it important to ask why we should conduct social science research in developing as compared to developed countries. One possible answer is that a variety of problems can be more severe (though not necessarily so) in developing rather than in developed countries, and that the means for addressing them (in other words, financial, institutional capacity, and so on) can be more limited in developing countries. Additional knowledge concerning these problems, generated through the collection and analysis of primary data, is often useful for understanding and formulating policy or institutional solutions. Furthermore, in some countries with less freedom to conduct research, outsiders can poke into social and political issues that would otherwise not be researched and, hence, challenge the status quo. In the best of all possible worlds, research effort should be directed in proportion to the severity of social, economic and environmental problems, though this is not always the case.

So – getting to our central question – why should we specifically do *field* research in poor developing countries? It is important for the following reasons:

1. Field research can reveal new or related problems that the researcher was unaware of. Researchers often go to the field with a preconceived idea of the scientific or social problem they want to address. Upon arrival to the field, they often discover that the problem of interest for the researcher is not the most urgent priority for people in the area (see Box 2.2).
2. Field research can serve as a vehicle for local people to comprehend and address a problem they are facing, thus making it possible to work towards a solution, or, at least, to give local people a means for inserting their views into the policy process (see Box 2.3).
3. Field research can be directed not only at understanding a problem, but also at monitoring and/or evaluating government policies and programmes that might have been put in place to address the problem. For example, field research can help to understand how integrated conservation and development projects (ICDPs) actually work, and to evaluate the real conditions that affect the success of those programmes. Other examples of programmes related to livelihoods in forested areas and environmental problems are: social and community forest programmes; eco-tourism; payments for environmental services (PES); and Reducing Emissions from Deforestation and Forest Degradation (REDD) schemes.

In keeping with the quotation from Bertolt Brecht that opens this chapter, the achievements of household field research can be justified even if they are modest and do not achieve 'infinite wisdom'. If the research can help lessen the effects of a problem by pointing out a policy error and leading to a policy course correction, it may end up having been worth the high costs involved and the disruption of daily life. Ultimately, the utility of field research to society is partly related to whose interests it serves.

Box 2.2 *Fieldwork as an eye-opener: An example from Guatemala*

José Pablo Prado-Córdova

Our research project was aimed at exploring the cause-and-effect relationships between the conservation status of the Guatemalan fir (*Abies guatemalensis* Rehder) and its socio-economic functions at the household level in the adjacent rural villages within this species' distribution area. Fieldwork was carried out during the period 2004–2006 and entailed interaction among villagers and botanists, plant ecologists, foresters, agronomists, entomologists, economists, enumerators and students from both the University of San Carlos in Guatemala and the University of Copenhagen. The original research question for the socio-economic component of this project dealt with estimating the economic importance of *Abies guatemalensis* in nearby peasant households. This was decided without consulting villagers regarding the extent to which this question was valid or even relevant for the proposed aim of the research project. Soon we came to realize that this species plays a minor role in local households' economies. We also learned that conservation threats such as poaching were more associated with external agents, who take the lion's share of the selling of illegally harvested branches, than with local agents. Fieldwork was an eye-opener for those involved in the project and made us adjust our original set of research questions in order to come up with a more realistic, problem-focused approach.

Box 2.3 *Participatory ethnocartography with the Achuar, Peru*

Martí Orta-Martínez

In a series of workshops held in Lima and Iquitos in 2005, the umbrella organization of the indigenous peoples in the Corrientes River (Federación de Comunidades Nativas del Río Corrientes, FECONACO) asked for research that mapped the activities of oil companies in their territory. FECONACO aimed to get scientific evidence for the environmental impacts that these activities have in the communal territory of the Achuar indigenous peoples.

To answer the call, a team of researchers of the Autonomous University of Barcelona (UAB) designed a Participatory Action Research (PAR) plan involving both UAB researchers and members of FECONACO. Researchers trained a team of indigenous monitors in the use of global positioning system (GPS) and digital and video cameras. After training, indigenous monitors walked the territory and collected information on old and new oil spills. Researchers cross-validated these data with results obtained from a temporal study of satellite images in order to assess the spatio-temporal environmental impact of the oil companies on the indigenous territory.

FECONACO has used the information generated by the team of researchers and indigenous monitors to initiate legal complaints to the government of Peru regarding the impact of oil companies in their territory. The research has empowered indigenous communities, allowing them to support their case in legal confrontations with the oil company. It has also raised the environmental standards of the oil company, with obvious environmental benefits for indigenous peoples.

Whose interests are served in doing field research?

The discussion above assumes there is only one frame of reference for judging the utility of field research: that of the academic community and the society at large. But of course there are various interests involved in weighing the merits of undertaking field research involving data collection through household surveys. In this section we focus the discussion on the interest of (a) society at large, (b) the community that is the target of the research, and (c) the person or team undertaking the research.

The interests of society at large

The discussion above basically justifies field research on the basis of increasing our scientific understanding of a given problem. We have also established that conducting field research is justified if it serves to understand, diminish and/or resolve the problem it is designed to address. But 'society at large' is a complex entity. Which part of society at large do we mean? Much of social science research in developing countries is funded by bilateral or multilateral donor organizations in collaboration with national governments and institutions. In the best of cases, all institutional parties that manage research are of one mind regarding the importance of the research. But, in some cases, the research is more a reflection of international rather than national priorities. Furthermore, national or more local priorities are not necessarily in agreement in some research interest areas. For example, national governments might not have an interest in research focusing on the social conditions of ethnic or religious minorities that outside researchers consider worthy of study.

Research regarding tiger conservation in India provides an example of how international agendas often dictate what needs to be researched. The rapid decrease of the tiger population has led both to policy responses by the national government and to an increase in research on the topic, but neither of these really take into account local priorities. Following an international trend (Smith and Wishnie, 2000; Chan et al, 2007), the government of India in 1973 enacted 'Project Tiger', a set of political measures to protect this emblematic species mainly through the creation of protected areas, such as tiger reserves and wildlife sanctuaries. Unfortunately, these measures generally do not take into consideration the presence of people living inside or around the protected areas. Research on the topic has focused on the biology of the species and on the causes of its disappearance, such as poaching or habitat destruction (Madhusudan and Karanth, 2002), but not on the interactions between local people and wildlife. Thus, policy measures and the research agenda have both followed trends established by international conservationist organizations and have neglected local priorities, such as access to natural resources, development and protection from wildlife. It is important for conscientious researchers to be mindful of these dynamics and of the fact that society is composed of many actors, not all sharing the same interests.

The interests of the community being researched

Does field research conducted in a given community end up serving the interests of that community? Ideally, this would be the outcome, though often it is not. And, even worse, bad field research can cause or aggravate problems in the community being studied.

Ideally, field research can at least provide an indirect benefit to a community by, for example, serving as the information base for development projects or policy reforms that eventually redound to the benefit of the community. In some cases, the benefit can be more direct – such as in cases where action research is focused on understanding and remedying only the problem experienced by the community. However, many field research projects fall short of these objectives. It is important for research institutions and individual researchers to attempt to design research in such a way that community interests are served, either directly or indirectly, in spite of the fact that this is a difficult challenge.

Frequently, respondents in target communities reap absolutely no benefit from research, in spite of having collectively put hundreds of unpaid hours into answering questions. This might result from any combination of bad preparation, poor design, implementation, data collection, analysis, policy outreach and policy impact, among other factors. A frequent retrospective lament of many university students and senior researchers is that their hard work in the field has ended up 'gathering dust on a shelf'.

Often, however, research is deliberately extractive in character and has no intention to directly benefit the community being researched. Often extractive research can be justified by indirect benefits to the community researched (for example, through policy change) and possibly to other communities to which the results may be generalizable. However, in some cases, field research does not attain even the indirect benefits sought.

How can we best assure that field research serves the interests of a given community, if that is a goal of the research? One way to do that is to involve local people in the conceptualization, design and/or implementation of the research (see Box 2.3). It is not always practical or possible to carry out field research in this way, but is an option that should be considered by researchers who are strongly inclined toward assuring that communities benefit from research efforts carried out in their midst. Another way of benefiting local communities is directly sharing the knowledge generated through the research with them, as was done by several PEN researchers (see Box 2.4).

Box 2.4 *Returning information to participants: The Community Baboon Sanctuary, Belize*

Miriam Wyman

A study was conducted within the Community Baboon Sanctuary (CBS), Belize, a small community-reserve under the World Conservation Union (IUCN) Category IV protected area status to protect the black howler monkey (*Aloutta pigra*). Research assessing conservation from different perspectives involved interviewing 135 of the approximately 220 landowners within the seven villages that make up the CBS. Additionally, fieldwork surveying forest and land cover change covered all seven villages. The research results were returned to the CBS villages through meetings and through dissemination of written materials:

1. Meetings

The researcher returned to the CBS to make a formal presentation to the Women's Conservation Group, the management body representing women leaders from the seven CBS villages that oversees conservation efforts and research within the CBS.

Additionally, the researcher visited each of the seven village leaders and organized a meeting in each of the villages for interested residents. The meetings provided a good opportunity to not only summarize the research findings, but also to answer the questions or concerns of residents regarding the goals and process of the research. Several meetings provided a forum for residents to communicate with each other on how this research could improve their livelihoods or resolve local management issues.

2. Written materials

Short, non-technical reports (3–5 pages) were developed and handed out to residents at all meetings and to each of the seven village leaders.

A laminated poster showing research results was used at every meeting and left with the director of the CBS. A copy of the dissertation was sent to the CBS director, as will copies of any future publications from this research.

Not only is sharing research results an ethical thing to do, but results can also help with future management decisions and support for future funding. In the case of this particular research site, the CBS director is interested in using this study's findings for future grant writing to improve conservation and development projects.

Field research can potentially uplift local people by valuing their knowledge. Many local people are used to having their opinions ignored. By trying to understand local perspectives and putting them in the public arena, researchers can provide an avenue of empowerment and communication between local people and authorities that otherwise would not exist. Finally, in some cases at least, especially where there is no research fatigue, local people might simply enjoy the interaction with somebody from the outside asking interesting questions, bringing pictures of faraway places and just spending time with them.

The interests of the researcher

Field research is done not just to meet societal and community objectives, but also those of researchers themselves. Various academic and personal interests motivate the implementation of field research by researchers and university students:

Academic interests: There is often a pedagogical component in social science curricula that gives attention to 'learning by doing' and learning by having first-hand contact with, and knowledge of, the day-to-day realities experienced by the people being studied.

Such curricula tend to espouse a training approach where the practicum embraces all aspects of the social scientific approach, from beginning to end: theory; methods; identification of a problem; formulation of a question or hypothesis; draft research instrument; pretesting; community household census; random selection of households; implementation of a household survey; data entry; data cleaning; data analysis; write-up; and restitution to the community. Indeed, the structure of this book reflects such an approach.

Relatedly, field research experience is sometimes a requirement for obtaining a degree in a specific field of study (such as anthropology or development studies).

Personal experience: Field research provides the opportunity for a unique personal experience. Field research entails the discovery of new places, challenging oneself, often learning how to work in a class, race, ethnic, cultural and linguistic context different from one's own and learning how to overcome difficulties in this unfamiliar context.

Understanding the world through field research is a valuable personal asset. In many types of employment (including jobs in the conservation and development sectors), field research experience adds to the value of an academic degree.

While a positive personal experience and growth are important motivations (and these help in doing good fieldwork), there is also a risk that fieldwork can become 'academic tourism'. It is important for researchers to reflect on their motivations for doing field research, and to answer the question: 'who will really benefit from this research?'

The challenges of field research

What has been stated above appears to be – we hope – a set of convincing arguments for conducting field research in general, and field research in developing countries in particular. As long as the interests of society at large, the community, and the researcher are all met, why would one hesitate to conduct such research? Reflecting on what happens in the real world of research, there are some good reasons to hesitate and to reflect deeply on whether field research is truly worth the investment. The main challenges in conducting accurate and useful field research are ethical and personal. We outline below some of the more important ethical challenges, and, in Chapter 9, our colleagues discuss personal challenges while being in the field.

Ethical challenges

Field research raises ethical issues and it is important to be fully aware of them beforehand to address them adequately. The following are among the most important issues that must be thought through before and during field research:

Perpetuating unequal power relations: Bear in mind that communities that are the subject of field research may not always have the power to authorize or object to the field research being done. They are often on the receiving end of a prior decision made by people in government and/or academia 'the research will be conducted in village X'. Even if researchers ask village or town leaders for permission to conduct research, there is often no latitude for the leaders to say no. Conversely, communities are often unable to promote (in other words, to fund) or authorize research in cases where they want it done (say, to reveal a problem), as villages often lack the financial means to fund research and authorities can prohibit research that is too sensitive.

Closely related is the fact that local people often have no say in the content of the research to be done. For example in the 1980s and 1990s, much of the content of social science research in forests in developing countries was motivated by concern for the protection of forests and biodiversity and gave little (or lesser) attention to the well-being and rights of forest dwellers (West et al, 2006; Chan et al, 2007). Similarly, since 2006–2007 the focus has shifted to the role of forests in the global carbon cycle. Participatory approaches (see Box 2.3) can address these problems to some degree. However, the participatory approach is not always an appropriate mode of research and it should not be used in situations where science is not well-served by the consultative process.

The unfortunate and often invisible reality is that among the three interest groups discussed above, the communities are the least powerful.

Releasing of sensitive information: During field research, sensitive private information might be disclosed, potentially resulting in embarrassment for or harm to community members. Researchers should protect the privacy of participants. To prevent the release of sensitive information in settings where interviews are conducted face to face, researchers should select their methods of data collection, processing and publication carefully. For example, researchers should strive to conduct household interviews privately – out of the hearing range of eavesdropping neighbours or local government officials – both to guarantee privacy and also to improve the quality of communication with (otherwise) reluctant respondents. Researchers can also minimize the risk of public release of sensitive information by assuring the confidentiality of the names of respondents and the names of research villages.

Issues related to data ownership: Researchers conducting field research assume they have full ownership of the primary data being collected. But in cases of research on sensitive topics, for example, on issues related to territorial rights or traditional ecological knowledge, local people might be interested in the use of the primary data for non-academic purposes. The matter of relinquishing control of research data enters into a realm of ethics that researchers seldom think about beforehand.

Unexpected outcomes: The implementation of research sometimes stirs up local emotions related to the problem being investigated. In those cases, government authorities might act against local people who have chosen to become vocal. Or sometimes, the published research leads to policy reform or rethinking of programme objectives, and these reforms or new objectives are sometimes contrary to the interests of people in the community. For example, documenting widespread illegal forest uses may lead to stronger law enforcement towards local violators. While the researcher might not have intended these outcomes, it does not alter the fact that – in the worst of cases – the outcomes can be devastating and cannot be changed.

Almost all research endeavours have some relation to politics and power, and for that reason alone, one must be mindful of the consequences of conducting field research.

Overcoming the challenges

The negative consequences described above can often be avoided with foresight and – above all – a conscientious attitude on the part of researchers. In various parts of this book, reference will be made to steps one must take to uphold ethical standards in designing and implementing a research project. Here, we merely list a few basic principles that should be borne in mind as researchers embark on their projects:

- Consider participatory research approaches, but only if they are feasible, practical, and are consistent with the topic being investigated (see Box 2.3).
- Make sure that target communities are adequately consulted prior to doing research, and engage in these consultations mindful of the unequal power relations mentioned above.
- If the research project involves indirect rather than direct benefits to the community, explain this candidly to community members.
- Guarantee anonymity in the processing and publication of data (for example, the names of respondents should not appear in publications) and then rigorously uphold the promise.
- Tell members of the community that you will give them a full accounting of what has been found through the research, and then come through on this promise. Returning information to communities can be done in simple and inexpensive ways (such as community workshops), even for graduate students with small budgets.
- Prepare yourself for the field. Minimize culture shock by getting a big head start in learning the local language and by informing yourself about customs, mores and traditions. The more prepared you are, the more enjoyable your field experience will be.

Conclusions

In summary, there are good reasons for doing field research and yet also various reasons to be hesitant before committing oneself to this time- and resource-demanding activity.

In deciding whether and how to do field research in developing countries, prospective researchers need to be mindful of whose interests are being served and the ethics of the research enterprise. Researchers need to go beyond the deceptively comforting assumption that a research project is well-conceived and worthwhile if it is initiated and funded by an international donor organization and endorsed by a national government. Researchers should be aware that their presence in the field, rather than contributing to the welfare of the people that supply the data and host them, can be potentially abused in power struggles and that research results can be used to fuel those struggles. So, in weighing the ethical considerations of field research, it is important to think through ways to avoid perpetuating unequal power relations and to affirm a moral commitment to the community by planning ways to guard sensitive information, by anticipating sensitive issues related to data ownership and by striving to conduct the research in such a way that unwanted outcomes are prevented.

In our elaboration of the challenges of field research in poor countries, we do not want to dissuade prospective researchers from undertaking this activity. Instead, our message is that field research can be of great service to the community being studied, to society at large and to the researcher if – and only if – serious and responsible thought is given to the challenges that surely lie ahead.

Key messages

- Field research has several benefits: it can be used to overcome a lack of data from existing sources, to understand the local context, to control data quality, and to open new frontiers of knowledge.
- To guard against unwanted outcomes, great care and forethought should be invested in understanding whose interests are being served: that of society at large, that of the population being researched, or that of the researcher.
- Fieldwork involves several ethical challenges: unequal power relations between outside entities (like the government, international donors and the researcher) and the community being researched; the need to guard against the release of sensitive information; field research data ownership; and avoidance of unwanted outcomes for the local population.

References

Bernard, H. R. (1995) *Research Methods in Anthropology: Qualitative and Quantitative Approaches*, Altamira Press, Walnut Creek, CA

Chan, K. M., Pringle, R. M., Ranganathan, J., Boggs, C. L., Chan, Y. L., Ehrlich, P. R., Haff, P. K., Heller, N. E., Al-Khafaji, K. and Macmynowski, D. P. (2007) 'When agendas collide: Human welfare and biological conservation', *Conservation Biology*, vol 21, no 1, pp59–68

Madhusudan, M. D. and Karanth, K. U. (2002) 'Local hunting and the conservation of large mammals in India', *Ambio*, vol 31, no 1, pp49–54

Malinowski, B. (1922) *Argonauts of the Western Pacific: An Account of Native Enterprise and Adventure in the Archipelagos of Melanesian New Guinea. Studies in Economics and Political Science*, Routledge and Kegan Paul, London

Pender, J. (1996) 'Discount rates and credit markets: Theory and evidence from rural India', *Journal of Development Economics*, vol 50, no 5, pp257–296

Smith, E. A. and Wishnie, M. (2000) 'Conservation and subsistence in small-scale societies', *Annual Review of Anthropology*, vol 29, pp493–524

Udry, C. (2003) 'Fieldwork, economic theory, and research on institutions in developing countries', *American Economic Review*, vol 93, no 2, pp107–111

West, P., Igoe, J. and Brockington, D. (2006) 'Parks and peoples: The social impacts of protected areas', *Annual Review of Anthropology*, vol 35, pp251–277

Chapter 3

Composing a Research Proposal

Arild Angelsen, Carsten Smith-Hall and Helle Overgaard Larsen

---

Science is facts; just as houses are made of stones, so is science made of facts; but a pile of stones is not a house and a collection of facts is not necessarily science.
Henri Poincare (1905, *La valeur de la science*, Flammarion, Paris)

What is a research proposal?

Research can be defined as 'a systematic investigation of a question or resolution, based on critical analysis of relevant evidence' (Walliman, 2005, p37). A research proposal is a concise presentation of the planned research, answering two key questions in particular: (a) *What* is the project going to investigate? And (b) *How* is the project going to undertake the investigation? The research proposal specifies the steps required to move from questions to answers by providing a logical, coherent and realistic plan of action.

A research proposal typically has two purposes. The *internal* purpose, which is the focus in this chapter, is to force the researchers to carefully design and plan the research project. The research plan should be sufficiently operational to provide the general structure for project implementation. Most projects involve several partners, and the research proposal creates a platform for the early planning and collaboration among them. The *external* purpose is to convince the reader that one has a good (important, relevant) research idea, that it is of high scientific quality (well-formulated, sound methods, competence of individuals involved) and that the planned project can be implemented (realistic time and resource use). External approval may be needed to obtain funding or research permits, to be enrolled in a PhD programme or to get new research partners; for an overview of external evaluation criteria see Peters (2003).

A wide range of publications focused on research processes and research designs is available (for example, Bechhofer and Paterson, 2000; Booth et al, 2008; Creswell, 2009). Many of these include advice on preparing research proposals. The process of developing a research proposal can be seen as consisting of eight steps. These are illustrated in Figure 3.1. The first step is to produce a good research idea, which is then refined into a number of objectives and research questions. Further consultation of the literature should lead to development of a conceptual framework and a set of testable hypotheses. This should, in turn, allow a specification of data needs, how these are to be collected and finally demonstrate how they are to be analysed. Box 3.1 provides the outline of a typical research proposal and Box 3.2 specifies four key indicators of the quality of a research proposal.

The initial five steps in Figure 3.1 answer the first basic question, namely *what* to investigate (and why). The next three steps focus on *how* to undertake the investigation and why the approach is appropriate. A good research proposal covers all the eight steps and has coherence among all elements. Although the steps are presented sequentially, developing a research proposal is an iterative process.

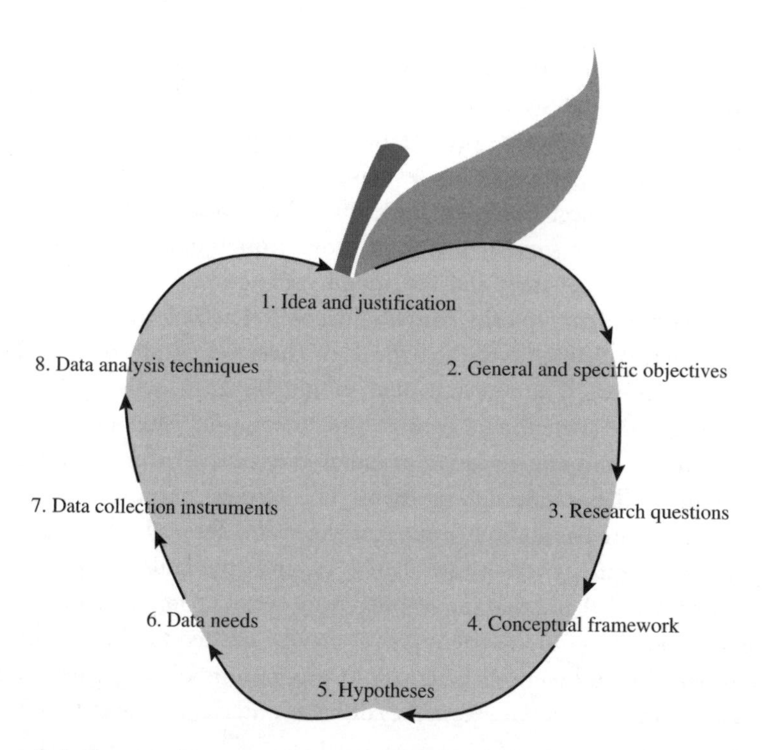

Figure 3.1 *Chewing the apple: Eight key bites in preparing a research proposal*

Box 3.1 *An outline of the research proposal*

A research proposal normally has the following headings (sections):

Title: Most researchers and graduate students use boring titles both on articles, papers and theses. Too bad, because people will then forget your work much more easily, or they may not even be attracted to read it in the first place. A good title is descriptive, of course, but one can also play with words. One possibility is to have a short catchy title followed by a more descriptive subtitle, for example, 'Ecuador goes bananas: Incremental technological change and forest loss' (Wunder, 2001). But be careful – one can go too far in 'clowning around'.

General objective: Presents the purpose of the research, including the (justified) problem leading to the need for the research. This should be brief, a maximum of ten lines. Here is one example, taken from a PEN proposal:

> The general objective of the proposed research project is to advance our understanding of the role of tropical forests in maintaining and improving rural livelihoods. Tropical forests are crucial to the livelihoods of millions of poor people worldwide. But just how important are they in preventing and reducing poverty? Which types of forests and products count most for the poor? Answers to such questions are essential to design effective forest policies and projects, and to incorporate forest issues in poverty reduction strategies. Yet we have surprisingly little empirically based knowledge to answer such questions adequately.

Background and justification: Most draft proposals have too much on this. Background is the easiest part to write, because one can learn a lot about most topics from the internet (please note, if used, remember to cite proper references). But set the scene: the geographical background, the policy debate and the state-of-the-art in the literature. And discuss why is it important to undertake this project – what is the rationale? What is the expected contribution to the policy debate, to the people in the area and/or the literature? This means that you should include a brief review of the relevant literature, with *brief* being the key word (again, most research proposals have too much of the literature review, and too little on what the project is going to do).

Conceptual framework (and theory model): This can include:

- a brief **review** of the relevant literature, major conclusions relevant for the work; and/or
- your **own framework** in terms of a box-and-arrow diagram on what the world looks like – at least the world relevant for your proposed project; and, in some cases,

- a **mathematical/analytical model** sketching the problem or outlining the hypotheses to be tested. But be careful: copying a textbook model into the proposal is rarely useful.

Specific objectives, research questions and hypotheses: Here the general objective is operationalized through formulation of specific objectives, research questions and hypotheses; the research matrix (Table 3.1) is a useful tool to ensure the logical relations among these.

Data collection: Specification of data needed to answer the research questions (the key variables), how to get them (collection), sampling (which groups, villages, and so on, to be included, how many households), and how to ensure data quality.

Data analysis: What specific methods will be used to analyse specific variables. The key word here is *specific*. Avoid a lengthy general description of methods, (almost) everybody knows what linear regression is.

Outputs: Specify what the planned scientific outputs are, such as articles, books and policy briefs.

Inputs: Present a list of all the input planned to go into the project implementation, including time (researchers, assistants), a budget covering fieldwork-related costs and institutional collaborators.

References: The literature cited (complete). Consider using reference management software.

And, by the way: **make it look nice!** No overkill, but show professionalism in the way the research proposal is written. For instance, use one font type and size consistently, consistent line spacing, numbered headings, a running title with page numbers in the headers, proper referencing in the text, and one consistent bibliographical style.

This chapter focuses on the first five steps. The next three steps are elaborated in Chapters 4–8 and 13. Developing the research proposal is part of the overall research process; note that not all components in the research process are covered by Figure 3.1, for example, data cleaning (Chapter 12) and communicating the research (Chapter 14).

A good research idea and the objectives

Each year, in universities around the world, the following scene unfolds. A student enters a professor's office, and asks the following question: 'I'm interested in topic X in country Y. Do you think I can write a thesis about it?' Topic X can be any number of things: poverty, environment, gender, food security, microcredit, forest use, aquaculture, fertilizer subsidies, trade liberalization or development aid. The list is virtually endless. Country Y, of course, could be one of the more than 100 developing countries. The student may receive a reply something like this: 'Yes, you can! In fact you can probably write about 100 theses about it. How many theses do you want to write?' The surprised student looks like a question mark and mumbles: 'Hmm, well, just one...'

What characterizes a good research idea? It should be original, related to theoretical developments in a particular field and/or practical problems that have received inadequate attention in the past (Sjöberg, 2003). It should address a research problem – the issue that leads to the need for the research (Creswell, 2009). Further, it should be possible to empirically investigate the idea within the temporal and financial constraints of the project.

So how do we come up with good research ideas? Starting points for generating good research ideas include the literature, attending and participating in scientific discussions (from the regular Wednesday lunch presentation to international conferences) and other events (such as seminars organised by non-governmental organizations (NGOs) working in your particular field) and getting into the field (directly observing issues and talking to people).

Most research ideas are initially too broad and too vague to be good or bad. For example, the relationship between poverty and the environment in Ethiopia is an interesting issue, but too broad for most research projects. Almost any research ideas can be made into good research proposals – or bad ones – although some ideas are certainly more challenging than others. How do we move forward and get from a possibly good idea to finding out whether an idea is good or bad? A first step in operationalizing the research idea could be to formulate the research objectives – what will the research attempt to achieve? The general objective (purpose) should be a broad formulation of how the proposed study feeds into the existing literature, for example, to contribute to the emerging body of knowledge on household and community-level climate change adaptation in developing countries. The specific objectives (the scope and number depends on the type of research proposal) should explicitly formulate what the study will contribute in the area specified in the general objective. For example: (a) develop a typology of adaptive responses to climate changes, based on empirical household-level data across a range of sites, and (b) explain the distribution of response types across households and communities. Good specific objectives are logically linked to systematically address the different components of the general objective, identified, for example, through literature review (Box 3.3). They use specific verbs, such as *determine, develop, calculate, compare,* and avoid vague verbs, such as *understand* and *study,* to convey the intended actions unambiguously. Further, the sequencing of specific objectives often move from the simple (quantification or determination of the problem) to the more complex analytical (explanation of causes and effects);

Box 3.2 *Four generic indicators of the quality of a research proposal*

There are several ways of assessing the quality of a research proposal, but four generic indicators are commonly accepted within the social sciences.

Construct validity refers to the degree to which the operational measures in a study reflect the theoretical constructs on which they are based – a survey questionnaire should use coherent and consistent measures of the variables of interest. For example, forest income must be defined so that it can be discerned from other income sources, such as income derived from trees on agricultural land.

Internal validity refers to the approximate truth about inferences regarding explanatory or causal relationships. It is, therefore, only relevant to research that seeks to make causal or explanatory inferences, but not to purely explorative and descriptive studies. Assume, for example, that a study aspires to investigate the impact of forest accessibility on households' dependence on forest income. This requires data that enable controls for other factors affecting forest income, such as market access, in order to establish internal validity. In general, internal validity is established by ruling out rival explanations to the hypothesized causal effect.

External validity refers to the ability to generalize. It is the degree to which the conclusions based upon a research study hold under other circumstances. The possibilities of assuring external validity are affected by the sampling approach (see Chapter 4), but also by the gathering of empirical evidence on context variables that can be used to situate the individual study in a larger empirical context (see Chapter 6).

Reliability refers to the quality of measurement, in other words, the consistency or repeatability of the measurements of variables. Reliability is pursued by seeking to minimize measurement errors in the implementation of the research (see Chapter 11).

Source: Trochim (2006); Yin (2009).

they must all be possible to realize given financial, temporal and other constraints.

Initially, graduate students and researchers typically generate many research ideas and, in the end, have to pick just one. Which one to choose among several good ideas? Three commonly used selection criteria are: First, what can be learned from this research? What is the potential for new insights? Second, one should consider choosing something that is policy relevant and useful to someone. Look through journals (popular and scientific), policy and other documents to find out 'what's hot and what's not'. Third, one should choose something that one is interested in and finds fascinating. That personal motivation is needed during the long evenings/nights that are required to complete the research project. While personal interest might be one good reason for the choice of research topic, Walliman (2005, p28) warns against making 'the choice of a problem an excuse to fill in gaps in your own knowledge'. Research should enlarge the public knowledge about a topic.

Research(able) question(s)

When the research idea and objectives are formulated and delimited, the next step is to develop one or a few research(able) question(s) in connection to each specific objective. The success of the research project depends a lot on how each research question is formulated. First, if the research questions are not well defined, this may result in the collection of data that will *not* be used and/or *not* collecting data that would be needed to answer the more specific research question eventually chosen. Second, a good research question helps limit the scope of the project. We are yet to see a research proposal that set out to do too little. Limitation can be achieved in many ways, for example, geographically (only Masindi district instead of Uganda), reducing units of analysis (only charcoal producers and not rural households) or stricter boundaries around the problem to be investigated (only one specific policy intervention, such as better policing along the roads for illegal charcoal producers). Third, a well-defined research question is an essential component in the structure of a research proposal. In particular, it helps to define the relevant literature, develop good hypotheses and define data needed.

Finding a good research question is often the most difficult part of the research planning. There is no simple recipe to follow, except: Read – discuss – think! Some proposals contain purely descriptive research questions, for example, 'what are the main income sources in the study area?' Getting an overview of the local economy may be important in the project but such descriptive questions make poor research questions. Research goes beyond simple accounting or finding correlations between variables – it also aims to understand and explain the world (for more on this, see Chapter 13). The research questions should reflect that. Desirable attributes of a good research question include:

- First, it is sharply defined in a way that can be answered in one sentence, possibly with 'yes' or 'no'. An example of a poor research question (although it can be used to formulate the general objective) is: 'Which factors influence the adoption of soil management practices?' A better one is: 'Does better access to formal credit lead to higher investments in terracing?'
- Second, it is *researchable* in the way that the proposed project can answer the

question, given particularly two constraints: (a) the time, money and skills available, and (b) the data availability and variability (enough variation in the data to test the hypotheses, see below).

- Third, a *puzzle or apparent paradox* makes a good starting point for making research questions. It might be a case with contradictory impacts, or contrary to conventional wisdom. For example, there is a research report from the highlands of Ethiopia suggesting that poor people invest *more* in soil conservation, contrary to what conventional wisdom suggests. Why? You formulate a research question based on this, and later make possible hypotheses to test (for example, the poor rely more on the land and therefore have stronger incentives, or soil conservation is labour intensive and the poor have low opportunity costs of labour).
- Fourth, it makes a *contribution* to our knowledge base. That means either that: (a) the research question is a new question or an old question reformulated and taken a step further, or (b) an old question applied to a new geographical area or a new (and better) set of data (to test the universal applicability of the answers proposed by earlier studies).
- Fifth, it is *policy relevant*, in other words, there are some policy handles and space for intervention (not necessarily by national government, but also by NGOs, local village council, firms or donors).

Conceptual framework

Having formulated objectives and research questions, the next logical step in the survey-oriented research process is to explicitly identify/develop and describe a conceptual framework. In practice, this is often done alongside the specification of objectives, research questions and hypotheses. A conceptual framework essentially defines and outlines relations between the main concepts a research project works with, based on selected, prevailing theories and associated empirical works. In other words, how the reality under study is expected to behave. A conceptual framework should provide guidance to the development of hypotheses and subsequent determination of data needs and selection of techniques for data analysis.

Conceptual frameworks come in many forms and vary across disciplines. Chapter 5 presents one that many students of poor rural economies have found useful, namely the Sustainable Livelihoods Framework (SLF, for example, Carney, 1998; Ellis, 2000). The SLF is a broad framework for analysing how household assets and higher scale factors (local institutions, markets, and so on) result in a set of livelihood strategies and associated livelihood outcomes (that have repercussions on household assets in the next period as well as on the

Box 3.3 *Characteristics of a good literature review*

Making a good literature review is challenging. Some literature reviews are like laundry lists: for example, Andersen (2002) said X, Brandon (2003) studied Y, Carlson (2004) looked at Z. These are poor reviews. A good review will structure the literature by topics and discuss it critically in a synthesis of results contributed and approaches applied by previous studies, ending with a conclusion stating what remains to be done in the field (this should then justify the relevance of the proposed study). For example, a good literature review on environmental dependence and poverty might start with a subsection on definitions and measurements of 'environmental dependence' and perhaps also 'poverty' (briefly, as there are numerous articles and books covering that). Next, it may have subsections on overall dependence, how dependence varies across household groups (rich–poor, female–male headed, and so on), and how dependence varies across different socio-economic environments (remote–central location, and so on). Finally, a section on what methodological approaches previous studies have used, their strengths and weaknesses in relation to studying environmental dependence and how your proposed research will expand the existing scholarship (for example, by including new relevant concepts, applying a different methodology or testing a hypothesized relation).

Providing a complete guide to conduct literature reviews is beyond the scope of this chapter and the reader is referred to the many available books (for example, Hartley, 2008; Creswell, 2009; Machi and McEnvoy, 2009) and online sources (for example, Obenzinger, 2005; Emerald, 2010; Melbourne, 2010).

environment). The broadness of the SLF is both its strength and weakness. It is useful as a framework for organizing information, as demonstrated well in Chapter 5, but it is less well-suited as a basis for development of explicit hypotheses – for this additional literature needs to be consulted.

The conceptual framework should not be confused with general theory development in the social sciences. A theory is by Walliman (2005, p439) defined as: 'A system of ideas based on interrelated concepts, definitions and propositions, with the purpose of explaining or predicting phenomena.' Thus, a major distinction between the conceptual framework and the theory is in terms of specificity. For example, the agricultural household models in the tradition of Singh et al (1986) can be seen as a specification of certain aspects of the SLF approach. While SLF gives a general link between markets and the choice of livelihood strategies, a simple agricultural household model might predict that higher off-farm wages will reduce the production of subsistence crops.

Depending on the purpose of the project (or the research proposal), the researcher may want to elaborate in detail the theory part in the proposal. This is, however, specific to the discipline and not discussed further.

Identifying, developing and describing a conceptual framework (and the associated objectives, questions and hypotheses) may be difficult and time-consuming. This should be done on the basis of an overview of the existing literature and identification of the research frontier in the field under study. Fortunately, good reviews are available in many fields and most research articles feature short literature reviews. A good point of departure is just to start reading and nesting (in other words, follow interesting references), for example, based on initial internet search (using bibliographic search engines or databases such as Google Scholar, ISI Web of Knowledge, Science Direct or Swetswise) and discussing ideas and draft texts with colleagues or fellow students.

Hypotheses

The next step is to develop hypotheses that should be formulated in response to each research question. A hypothesis is a reasonable scientific statement that is put forward for empirical testing. Beveridge (1950) refers to it as the principal intellectual instrument in research.

The difference between the research question and hypotheses is not always clear-cut. It depends on how they are formulated. We often recommend the following approach and balance between the three first components of the proposal: one grand research idea, 1–3 specific objectives, a few (maximum 4–5) research questions, at least one and maximum 3–4 hypotheses linked to each research question. But, there is a limit on the total number of hypotheses (not more than 6–8).

There are two principal ways of deriving the hypotheses, corresponding to two fundamentally different scientific approaches: (a) the *deductive* (or hypothetico-deductive) approach, where hypotheses are derived from theoretical models, and (b) the *inductive* approach, where hypotheses are based on empirical research findings (Creswell, 2009). In many proposals, the hypotheses will be developed using a combination of the two approaches (and theories are often being modified to accommodate new empirical results). In the research method discussed in this book, surveys and hypotheses should be based on both a consistent conceptual framework (and theories) as well as a review of the existing empirical knowledge. This will make it possible to directly relate research findings to larger scientific discussions. A common flaw, for example, in Master theses, is 'ad hoc reasoning' where results are *not* presented or discussed in a larger theoretical context.

Desirable attributes of a good hypothesis include that it leads to new knowledge and that it is testable. A hypothesis can lead to new knowledge by taking well-established facts further through combining them and thereby creating (and testing) new links between variables. Consider this example:

Fact 1: Many farmers are credit constrained.

Fact 2: Soil conservation often requires cash investments.

Hypothesis: Credit constraints discourage farmers from investing in soil conservation.

This hypothesis combines the two general facts, yet there are many reasons why this might not be true: other constraints can be more important, or farmers give soil conservation investments low priority and make other investments when credit supply increases. Trivial hypotheses that do not generate new knowledge should be avoided. For instance, the hypothesis 'household consumption increases with higher household income' is based on theory and can be tested with appropriate data, but is not very interesting as the relationship between income and consumption is already well-established.

An absolute requirement to a hypothesis is that it is testable. The formulation, in combination with data to be generated by the proposed research (next section), must ensure this. Hypotheses derived from mathematical models, with fewer variables and more precise predictions regarding relationships, are normally more specific and testable.

Consider the following hypothesis: Households' dependence on environmental income depends on household and village characteristics. This is poorly formulated because it is too general, for example, it does not specify which characteristics influence environmental dependence in what way. Thus a better hypothesis would be: Households in geographically remote areas have higher environmental dependence. This hypothesis seeks to establish the correlation between two specific and measureable variables. This might be taken a step further, by asking the 'why' question (Chapter 13) and then formulating hypotheses that test the reasons why environmental dependence is higher in remote areas (assuming that is the case). The reasons could be low prices of agricultural commodities, few off-farm work opportunities, better access to forest resources, and so on, and associated hypotheses would be of the type: Poor access to agricultural markets leads to higher environmental dependence.

All variables used in hypotheses must be measurable, in other words, they should be operationally defined. For instance, the reformulated hypothesis should be accompanied by text defining the four key terms used and describing how they can be measured: households (for example, group of people living under the same roof and pooling labour and income resources); remote areas (for example, using the proxy of travel time to nearest major market or the district capital); dependence (two typical measures are absolute income, and relative income, in other words, share of total household income); and environmental income (see, for example, Sjaastad et al, 2005). This specification may result in the need for definition of additional terms (for example, what is meant by labour pooling), until it is unambiguously defined at the level where each variable can be measured in the field.

Data needs

Having completed the first part of the research proposal (the first five steps in Figure 3.1, related to 'what to investigate'), the next stage is to specify data needs, how data are to be collected and finally how they are to be analysed (the last three steps, related to 'how to investigate'). A useful tool to ensure consistency between the different steps in the research process is the research matrix (exemplified in Table 3.1). The research matrix is simply a set of columns of four of the eight elements in the research process outlined in Figure 3.1. It can be expanded to include data collection instruments, conceptual/theoretical underpinnings or other non-research process items, such as a column on policy relevance. A well-prepared research matrix will ensure a robust research design with strong, consistent and clear relations from research questions and hypotheses to methods for data collection and analysis. The process of preparing a research matrix often leads to modification and delimiting of research questions and hypotheses, as it becomes clear that data requirements are too large or that research questions are ambiguous.

The data needed depend entirely on the hypotheses to be tested. Here we want to highlight some of the generic and critical issues in the process: the population from which the data are being collected and the variables to be included in the data collection. The former concerns the site selection and sampling strategies discussed in Chapter 4, while the latter concerns the choice of survey types and design of survey tools discussed in Chapters 5–7.

When considering the population to study and the variables to elicit, it is critical to ensure that the necessary variation is present in the data. Consider the research question: 'how does charcoal production vary with the charcoal price?' The price response cannot be estimated by a one-shot household survey from three neighbouring villages, as the variation in charcoal prices at one point in time within a limited geographical area is likely to be small. There is simply not going to be sufficient price variation to permit testing the hypotheses derived from this question. Thus a key principle is: Design your research so as to get variation in the variables applied for testing the hypotheses, and keep the other variables constant. A (random) sample of households in three neighbouring villages will provide natural variation for studying, for example,

Table 3.1 *An example of the research matrix in relation to the specific objective 'to analyse the roles of forest resources in rural livelihood strategies'*

Research question	Hypotheses	Data needed	Methods of data analysis
1. What is the share of forest-derived income in rural households?	1.1 Poorer households' forest-derived income is higher relative to total income while richer households' forest-derived income is larger in absolute measures.	Total household income.	Forest-derived income per income quartiles.
		Forest-derived income.	Chi-square testing – comparison of conditional means.
2. Do natural forest resources provide important insurance values in coping strategies?	2.1 In situations of common and idiosyncratic income shocks, forest resources have an important insurance function.	Household-level recording of (a) type of shocks, (b) impact of shock, and (c) responses to each shock type.	The importance of forest-based and other responses across shock types and income quartiles – comparison of conditional means.
	2.2 Forest-derived income does not provide sufficient means to smooth consumption relative to the risk/shock faced by households.	Total household income. Forest-derived household income. Consumption survey.	Table analysis of forest-derived income and consumption per income quartiles. Chi-square testing.

Source: Adapted from Nielsen (2006)

the role of forest extraction over the life cycle of a household, where the varying variable could be age of household head. Other household characteristics with natural variation would be income (poverty), assets, household headship, household size, and so on.

Variables such as prices and institutional arrangements show less variation across households in a community, and a study of these will require careful selection of study area and communities/villages to be included to get the required variation. Market accessibility (central–remote location) is often a key gradient in livelihood strategies (Ellis, 2000) and is often positively correlated with factors such as population density and forest abundance. Interesting research questions that use this dimension can be developed. On the other hand, for research not related to the location continuum, variation along this gradient is best avoided. To illustrate this point, consider a comparative study of local institutions in terms of the poor's access to the forest: Having selected villages with different institutional arrangements, if these also vary in terms of market access – and market access and institutions are correlated – how can one be sure that the differences observed are due to the different institutional arrangements and not market access?

Some research topics require time series data. Indeed, a large set of new research opportunities opens up with panel data and access to earlier surveys (with household contact information) will enable construction of panel data sets useful for, for example, testing or exploring how environmental dependence varies with types of household trajectories. The study of poverty dynamics, for example, has been greatly facilitated by the better availability of panel data (Deaton, 1997; Carter and Barrett, 2006). Even without household contact details, access to previous survey results may allow for studies on village characteristic trends or provide important contextual background information.

Data collection and analysis

Once the idea and the conceptual framework of the research proposal have been developed; the objectives, research questions and hypotheses defined; and the data needs identified, the laborious tasks of setting up and implementing the data collection instrument – the questionnaire – awaits. This is no trivial matter, as the questionnaire needs to elicit information on all variables necessary for assuring internal validity (in other words, conducting the analyses required to answer the research questions) in a way that does not compromise construct validity (in other words, actually measuring the theoretical concepts outlined in the research questions) (see Box 3.2). Keep in mind, however, that the quality of data arising from even the best questionnaire design can be seriously compromised by faulty data collection. Pitfalls and opportunities of designing questionnaires are presented in Chapter 7, while Chapters 9–11 are devoted to strategies ensuring high quality data during fieldwork.

Data analysis starts when (at least some) data have been collected. But methods for analysing data need to be decided when the research questions are formulated to ensure that data on all required variables are available for research to go beyond descriptions and basic correlations (as well as for proper sampling strategies). Chapter 13 gives an introduction to data analysis.

We stress the need to continue the iterative research design process during data collection. This entails scanning data collected during the pilot test of the questionnaire for detection of ambiguities and keeping an eye on the data collected with the final questionnaire. It is valuable to monitor the collected data in terms of, for example, missing or unreliable data on key variables that could endanger the data analysis. Performing basic correlations and cross-checks on the preliminary data may provide hints for follow-up during collection of contextual data.

Conclusions – what can go wrong

Developing the research proposal is perhaps the most critical stage in the research process, with ample opportunities to commit irreparable mistakes. Some of the most common, and often fatal (for example, in terms of getting funding), are:

- **Lack of clear justification**: What will be the contribution of the research project? Structure the proposal in a way that conveys and justifies the purpose and importance of the proposed research. This includes succinctly outlining: (a) the problem, (b) to whom it is a problem, and (c) what consequences of filling the knowledge gap can be foreseen. This can be done by stating the problem and the general objective at the very beginning of the proposal and providing justification for the existence and importance of the problem in the background section.
- **Lack of in-depth critical literature review**: A poor literature review is characterized by (a) focus on spatially limited references, such as only those relevant to the country where empirical work is planned to take place, (b) lack of consideration of non-sectoral knowledge, for example, a study on common forest resource management could benefit from studies on common pasture or water management, (c) failure to locate key studies and thus the research frontier, (d) emphasis on determining answers rather than identifying questions. A good literature review should result in identification of research gaps and the formulation of a relevant general objective, specific objectives and associated research questions.
- **Too broad/vague objectives, research questions and hypotheses**: Specific objectives should be operational. Research questions and hypotheses need to specify the variables for observation/measurement/analysis and nature of relations/influences. It is advisable to always discuss specific objectives first with peers and supervisors, followed by research questions and hypotheses, before proceeding to specify data needs and choosing methods. This can save a lot of iteration and time.

- **Lack of coherence**: Many proposals do not maintain a chain of logic from the identified research problem through formulation of objectives and research questions to choice of data analysis methods. Potential pitfalls are manifold, for instance, (a) during proposal development the focus is gradually moved from the research questions to other issues, or (b) the proposed data collection does not suffice for testing of the proposed hypotheses due to too few observations or lack of variation in the data.
- **Weak method description**: The reader should get a clear idea of how data will be generated (the surveys, selection criteria for study sites, villages and households, sampling strategy, number of households to be interviewed, questionnaire components) and the methods for analysing the data. Try to become sufficiently familiar with methods to enable their operational integration into proposals and get peer and supervisor comments before finalizing the proposal. Also take the acid test of clarity using your inner eye: when reading your method section can you see clearly, step by step, how required data is generated? If something is missing or unclear, revise the text.
- **Poor quality of writing**: Poor language indicates poor thinking (and ability to implement the research project) and it is not accepted as an excuse that many of us are writing in our second or third language. This can be addressed through courses in academic writing and, best of all, practice. Share draft proposals with peers and supervisors and make sure you respond to critical comments. If necessary, pay a professional copy editor to go through the proposal.

Key messages

Application of an iterative research design process (Figure 3.1), using the research matrix (Table 3.1) and considering the four indicators of the quality of a research proposal (Box 3.2) will minimize incoherence between objectives, research questions, data variables and analyses. A good research proposal is characterized by:

- Addressing an important issue in a novel way.
- Containing information about all the eight steps outlined in Figure 3.1 to answer the 'what' and 'how' questions.
- Having consistency and coherence between the eight steps in the research proposal.
- Being realistic to implement within the time and budget available.

References

Bechhofer, F. and Paterson, L. (2000) *Principles of Research Design in the Social Sciences*, Routledge, London

Beveridge, W. I. B. (1950) *The Art of Scientific Investigation*, William Heinemann, London

Booth, W. C., Colomb, G. G. and Williams, J. M. (2008) *The Craft of Research*, third edition, University of Chicago Press, Chicago, IL

Carney, D. (1998) *Sustainable Rural Livelihoods: What Contribution Can We Make?*, paper presented at the DFID Natural Resources Advisers' Conference, July 1998. Department for International Development, London

Carter, M. and Barrett, C. (2006) 'The economics of poverty traps and persistent poverty: An asset-based approach', *Journal of Development Studies*, vol 42, no 2, pp178–199

Creswell, J. W. (2009) *Research Design: Qualitative, Quantitative, and Mixed Methods Approaches*, third edition, Sage Publications, London

Deaton, A. (1997) *The Analysis of Household Surveys: A Microeconomic Approach to Development Policy*, The Johns Hopkins University Press, Baltimore, MD

Ellis, F. (2000) *Rural Livelihoods and Diversity in Developing Countries*, Oxford University Press, Oxford

Emerald (2010) 'How to ... write a literature review', Emerald Group Publishing, http://info.emeraldinsight.com/authors/guides/literature.htm, accessed 9 April 2010

Hartley, J. (2008) *Academic Writing and Publishing: A Practical Guide*, Routledge, London

Machi, L. A. and McEnvoy, B. T. (2009) *The Literature Review*, Corwin Press, Thousand Oaks, CA

Melbourne (2010) 'Literature reviews', The University of Melbourne, http://unimelb.libguides.com/content.php?pid=87165&sid=648279, accessed 9 April 2010

Nielsen, Ø. J. (2006) 'Welfare consequences of deforestation and forest degradation in Mozambique', unpublished PhD research project proposal, Faculty of Life Sciences, University of Copenhagen, Copenhagen

Obenzinger, H. (2005) 'What can a literature review do for me? How to research, write, and survive a literature review', Stanford University, http://ual.stanford.edu/pdf/uar_literaturereviewhandout.pdf, accessed 9 April 2010

Peters, A. D. (2003) *Winning Research Funding*, Gower, Hampshire

Singh, I., Squire, L. and Strauss, J. (1986) *Agricultural Household Models: Extensions, Applications, and Policy*, World Bank, Washington, DC

Sjaastad, E., Angelsen, A., Vedeld, P. and Bojö, J. (2005) 'What is environmental income?', *Ecological Economics*, vol 55, pp37–46

Sjöberg, L. (2003) 'Good and not-so-good ideas in research: A tutorial in idea assessment and generation', SSE/EFI Working Paper Series in Business Administration No 2003:12, Stockholm School of Economics, Stockholm

Trochim, W. M. K. (2006) 'Research methods knowledge base', www.socialresearchmethods.net/kb/index.php, accessed 31 August 2010

Walliman, N. S. R. (2005) *Your Research Project: A Step-by-step Guide for the First-time Researcher*, Sage, London

Wunder, S. (2001) 'Ecuador goes bananas: Incremental technological change and forest loss', in Angelsen, A. and Kaimowitz, D. (eds) *Agricultural Technology and Tropical Deforestation*, CAB International, Wallingford, pp167–195

Yin, R. K. (2009) *Case Study Research: Design and Methods*, Applied Social Research Methods Series vol 5, Sage, London

Chapter 4

Sampling: Who, How and How Many?

Gerald Shively

The universe extends beyond the mind of man, and is more complex than the small sample one can study.
Kenneth Lee Pike (1962, *With Heart and Mind*, Eerdmans, Grand Rapids, Michigan)

Introduction

Research questions and hypotheses inherently provide information on data needs, in terms of what variables should be measured and with what variation (see Chapter 3). Collecting data with the desired properties requires one to:

- define a target population and sample;
- decide how to draw the sample; and
- identify a sample size.

As with most aspects of empirical research, data collection requires that theoretical concerns (of which there are many) are weighed against practical considerations. Statisticians have developed elaborate rules, guidelines and formulae for developing sampling strategies. This formal approach to sampling is important, but in many cases individual researchers working with small budgets and narrowly defined goals will find that practical concerns carry the day. In most cases small samples – say those of 1000 respondents or fewer – are unlikely to meet many of the objective criteria for sampling laid down by statisticians. Nevertheless, many important research questions can be investigated with small samples provided the samples are drawn with care and their strengths and weaknesses are well understood.

This chapter aims to assist in thinking through a range of issues associated with sampling by providing a brief overview of the common challenges and

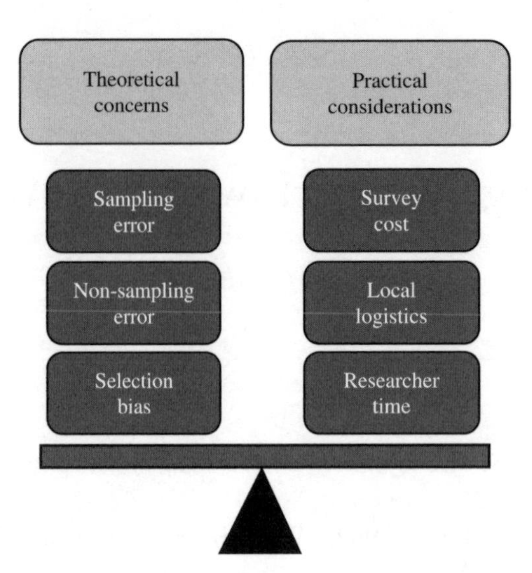

Figure 4.1 *Balancing theoretical concerns and practical considerations*

decisions regarding sampling (see Box 4.1 for an overview of sampling terminology). The focus will be on the issue of sampling in the context of quantitative research involving the systematic collection and analysis of standardized numerical data regarding individuals, households and their activities. It is important to point out, however, that sampling is an equally important issue in the context of qualitative research, especially in the selection and construction of case studies that provide detailed narratives that describe and explain observed patterns.

Box 4.1 *Sampling terminology*

A **sampling strategy** combines a set of procedures to select a sub-group of observations from the target population with a set of procedures to subsequently estimate some feature of that larger target population based on knowledge gained about the sub-group.

The **target population** is the group being studied, in other words, the group about which one wishes to draw conclusions.

The **sample frame** is the list of those elements or individual observations in the target population from which the sample will be drawn.

The **sample** is the sub-group of the target population actually studied, in other words, the group from which data are collected.

The **individual observations** within the sample (which could be households or individuals) are called sample elements or simply sampling units.

Who: Defining the target population and sample

Defining research questions and hypotheses (Chapter 3) as well as a target population are the necessary first steps in developing a sampling strategy. When defining the target population, care is required. On the one hand, if the target population is defined too broadly (for example, all rural households), resource constraints may not permit one to draw a sample that fully represents the target population. On the other hand, if the target population is defined too narrowly (for example, teenage boys who fish on Saturdays in the village of Easterling), the results may not yield answers that are broadly applicable, generalizable or interesting to others. However, even in the case of a meaningful but narrowly defined target population, the individual researcher is unlikely to have either the time or the resources to collect data from everyone in the target population. In this case, a procedure must be followed for choosing survey respondents. A sampling strategy combines a set of procedures to select a sub-group of observations from the target population with a set of procedures to subsequently estimate some feature of that larger target population based on knowledge gained about the sub-group. The sub-group is called the sample and individual sample observations are called elements, for example, villages, households or individuals. Provided one can identify and collect a sample that is fully representative of the target population, one can generalize about that target population based on findings from the sample. In this sense, sampling is an efficient alternative to undertaking a full census of a target population. At the outset of drawing a sample, therefore, one should think carefully about the target population of interest and the sampling strategy that is likely to be the most useful for drawing conclusions or inferences about that population.

As indicated above, the first step to be taken when starting any survey is to define the relevant research questions and hypotheses as well as a target population of interest. Note, however, that defining a target population in terms of activities may be problematic depending on the objective of the research. For example, for a study that aims to explain patterns of environmental dependence, the relevant population of interest (all who depend on the environment) cannot be known *ex ante*, that is before patterns of dependence have been empirically explored. More importantly, if the research aims to explain why some individuals participate in certain activities, the sample must include not only those individuals who actually engage in the activity of interest, but also a comparison group of non-participants. In such a case, a useful target population would be those individuals who live in an area with access to forests. The practical implications of this can be illustrated with a simple example. Consider embarking on research in a village of 300 households, with the goal of learning who in the village collects forest products and what factors are associated with this livelihood strategy. It would be difficult to understand why some individuals collect forest products unless the sample includes individuals with similar forest access opportunities who differ in terms of collection activity. In the case of our hypothetical village, if it is the case that most people in the village do not collect forest products, then one must be sure to collect information from enough residents to generate a meaningful sample of those who do. If only one in ten residents collects forest products, and a random sample of 30 households is drawn from the village, then one might reasonably expect a sample containing three forest users and 27 non-users. In this case, 27 observations may be enough to learn something about non-users, but three observations may be insufficient to feel confident about one's knowledge of forest users. The obvious solution to this dilemma is to draw a larger sample (so as to get a larger number of forest users). On the other hand, if everyone in the village collects forest products, no village sample (including complete enumeration) will tell us anything about non-users. In that case, the sample must be expanded to include other villages.

In some cases, conclusions based on a sample may not provide accurate information about the target population of interest. In other words, the sample estimate of a variable may be an inaccurate estimate of the true value of that variable for the population of interest. This situation is more generally referred to as systematic or non-sampling error. In such cases, conclusions can be drawn about the sample, but not the target population. Understandably, non-sampling error may be very hard to detect and can only be prevented by ensuring that the sample is drawn in such a way that it captures all the important characteristics of the target population.

Even a perfectly randomly selected sample does, however, not by itself guarantee that the sample captures all important characteristics of the target population from which it is drawn. Samples are draws from a population, just as a hand of five cards is a draw from a deck of 52. Some draws are 'luckier' than others. Unlucky draws are those that miss some types of individuals or households. Statisticians call this sampling error: the sample provides too many observations of some kinds and not enough observations of others. Sampling error is more likely to be a problem in small samples drawn from heterogeneous populations (as in the case of the three forest users in the example above) and less likely to be a problem in large samples drawn from homogeneous populations. It may be the case that a sampling approach that lends itself to provision of an unbiased estimate of a population parameter nevertheless results in a sample that is too small to provide a precise measure of that variable's true value in the population (Byrne, 2002). Sampling error can usually be avoided *ex ante* by increasing the size of the sample. What if you are stuck with a sample in which sampling error may be a problem? There are several econometric strategies to deal with sampling error, including bootstrapping, an approach in which one draws multiple random samples from a data set, so as to construct a new data set that is believed to closely approximate the actual distribution from which the sample was drawn (Hoyle, 1999). This is not a sure-fire fix, however. Moreover, non-sampling error can only be eliminated by improving the design of the survey, to increase the likelihood that one captures all aspects of the target population.

The necessary stages in sampling design

A sampling strategy should ensure that appropriate data are collected at each level relevant to one's analysis. If done properly, this means that data from different levels can be brought to bear in understanding the social, political and physical environments in which households operate. Too often, at the analysis stage of a study, a researcher must use a single binary indicator, for example a village dummy variable, to account for important differences among groups of households in the sample, such as the degree of market access, the strength of village leadership and the local burden of communicable disease. Proper collection of relevant data at the village level, therefore, may be just as important as the collection of individual household data (see Chapter 6). In general, sampling is pursued in several stages (Byrne, 2002):

Stage 1: Selection of the study area (district or region)

Most studies will be restricted to one or two region(s) for practical and cost reasons, but including households from a larger area (several regions) may be needed in some cases to obtain sufficient variation in the data or for doing comparative analysis and impact assessments.

Step 2: Selection of the geographical sub-units (villages)

Normally one cannot include all villages from a district or region in the study area, and therefore some degree of village selection is necessary. Selection should be undertaken with two considerations in mind: (a) including villages that are representative for the study area, and (b) providing variation across the villages in key variables to be studied. In some cases, one must decide whether to pursue relatively few households per village and a large number of villages, or instead concentrate on a large number of households in a few villages. The former option is likely to capture more variation in the data, and that is often a desirable feature because without meaningful variation in the data, there is nothing to study! If it is important to be able to compare villages, a rule of thumb is that a minimum sample size of 25–30 households from each village is appropriate for a village that ranges in size from 100 to 500 families (see more on sample size below).

Step 3: Selection of households within chosen geographical sub-units

Approaches and practical procedures for selecting households are highlighted further below. From a purely statistical point of view, random selection of households will be necessary to obtain a representative sample and thereby permit one to draw conclusions regarding the full target population from the sample statistics. However, randomness alone is not sufficient, especially in small samples, for the reasons highlighted previously.

The steps outlined above are operationalized through the use of a sample frame. A sample frame is a list of those in the population from which the sample will be drawn. The sample frame may be developed by conducting a complete enumeration of a survey area prior to conducting the survey, or it may be based on some other comprehensive list provided by authorities (such as voter rolls) or by others. If old lists are being used, it is essential that efforts are undertaken to update them before drawing a sample.

Where lists of names are not available, it is sometimes possible to develop a sample frame on the basis of a detailed map of a geographic area. Ideally the sample frame should accurately represent the population of interest. For this reason, a researcher must be very attentive to the problems of overcoverage and undercoverage. Overcoverage occurs when the sampling frame includes elements that are not part of the target population. For example, if a survey is intended to cover the population of working-age men and women, but the sample is drawn from all adult members of a community, some data may correspond to elderly individuals who no longer work. Including those data points in the analysis could generate biased estimates of parameters for the population of interest. A similar set of problems could arise if the most complete list of village members comes from a list of eligible voters during the last election: if the list is based on individuals, large households with many adult members may have a disproportionately high probability of being included. Fortunately, overcoverage is usually easy to detect and correct in the field.

Undercoverage is a more pernicious and difficult problem to detect. It arises when the sampling frame excludes elements that actually occur in the population. In this case, if observations are drawn from the sample frame, it will never be possible to collect information from the excluded elements. In the example of the list of eligible voters mentioned above, undercoverage may arise if the list excludes recent migrants or newly established households. Another example would be constructing the sample frame from a list of attendees at an extension training seminar. If many households did not participate in the training, and the reasons for non-participation are systematic within the sample, then undercoverage can lead to severe bias, typically of an unknown and hidden nature.

How: The mechanics of sampling

The multiple methods of drawing a sample from a population can be grouped into probability and non-probability methods. Probability sampling refers to a method of sampling in which the researcher can determine the statistical likelihood of selecting a unit of the target population into the sample. The researcher chooses to study individual units from the population. These sampled units are used to accurately represent the target population as a whole. The most common way these sample units are selected is by randomization. Random selection ensures that each unit in the population has a non-zero (and measurable) chance of appearing in the sample. The driving logic behind random selection is that the features of a random sample should match those of the population, and that a large series of random draws from the population is most likely to result in a data set that resembles the underlying population in important respects. A number of approaches fall under the heading of randomization, among them simple random sampling, systematic random sampling, stratified random sampling and clustered sampling. These techniques are outlined in detail below. All rely on the proposition that probabilities of selection into the sample can be used to recover accurate estimates of the true population parameters. This accuracy can only be assured if probability sampling is used at each and every stage of the sample selection process.

Non-probability sampling, on the other hand, refers to a method of data collection that does not rely on randomization and does not purport to deliver statistically accurate estimates of population parameters. No statistical theory can be invoked to guide the use of this method. In non-probability sampling, the odds that specific units of the population will be selected as part of the sample are generally not known and are not measurable across the underlying population. Non-probability sampling may be used when probability sampling is unnecessary, infeasible or too burdensome to carry out. In some instances, the researcher may not be interested in generalizing findings to a larger population or may not have the financial resources to facilitate probability sampling. In still other cases, probability sampling may be ruled out because of the specific nature of the sample or because the researcher has heuristic objectives (Tuli and Chaudhary, 2008). For example, in some cases, a researcher may be simultaneously developing a line of inquiry and a sample. In this situation, additional categories for sampling may only become apparent, and added, during the course of investigation (Sen and Sharma, 2008). Among the techniques available for non-probability sampling are convenience sampling, purposive sampling, contact or snowball sampling, and quota sampling.

In some instances, probability and non-probability methods may be combined, such as when a specific type of individual or group is purposively identified and then sampling units are chosen randomly from within that group. So, for example, if one wants to learn whether migrants differ from non-migrants in their use of forest resources, one might purposively select two specific locations for study with the knowledge that one site has been settled by migrants and one has not. Then, within each location, samples might be drawn at random. In most cases, non-probability sampling results from situations in which the researcher intends to study a population that is small, unique or difficult to find. An important issue to keep in mind, however, is that non-probability sampling can lead to bias vis-à-vis estimates for the underlying population, and this bias is typically of an unknown direction and magnitude. In the next section, the most common probability and non-probability sampling methods are briefly outlined.

Probability sampling methods

Simple random sampling

A simple random sample is just that: a group of respondents chosen simply at random. In a village of 400 households, for example, one might draw a 25 per cent sample by selecting 100 households for interviews. The process sounds straightforward, but behind this simplicity lies a requirement that one be thorough and not cut corners. There are two obvious ways to generate a proper equal-probability random sample of elements.

One option is to start with a reliable list of households in the village (for example, from a census or election). The next step would then be to make certain that the list is complete, fully inclusive and up to date. This could be accomplished, for example, by presenting the list to some key informants (such as village elders, local extension personnel, a religious leader) and adding or deleting households from the list as appropriate. With a comprehensive and accurate list in hand, one then numbers the households and chooses among them randomly, perhaps using a random number generator to select households for inclusion. Most statistical software packages and even some calculators can be used to generate random numbers. Alternatively, one can write names or numbers of all elements of the sample frame on strips of paper or cards and draw these at random from a box, without replacement, somewhat like a lottery. A more sophisticated approach is to use a 20-sided die for the task. Bethlehem (2009) describes some clever tricks for working with a die or other randomizers.

A second approach is to use aerial photos and number all the houses in the village. One then draws random numbers and selects the sample on this basis. This, of course, requires that high quality aerial photos are readily available for the task. Whatever method of random sampling is chosen, keep in mind the definition of simple random sampling: each unit should have the same probability of being selected. Note, however, that while equal-probability sampling is conceptually straightforward and the easiest probability-based method to implement, other sampling designs can result in equally precise, or even more precise statistical estimators (Bethlehem, 2009).

Systematic sampling

Systematic sampling shares features with random sampling, but selection at random is replaced by selection following a systematic rule that produces the same effect. So, for example, if one is aiming to achieve a sample of size s from a sample frame of size n, one begins by choosing a step length $a = n/s$, where a is understood to be the smallest integer next to n/s. A starting point between zero and a is picked at random and then every ath household is selected from a randomly prepared list of the sample frame. As an example, to pick 30 households from a village roster of 280, $s = 30$, $n = 280$ and $a = 9$. Starting randomly with the third household on the roster, one would pick households 3, 12, 21, 30, 39, and so on. It must be emphasized that the $a = n/s$ approach must be applied to a randomly prepared list of the sample frame. If one applies the method to, say, an alphabetically sorted village register, the rule will not produce the desired effect and the sample will not be randomly drawn.

Stratified and clustered sampling

Stratified sampling is a method used when the researcher has some a priori understanding of particular characteristics of the group of interest and posits a relationship between these characteristics and one or more dependent variables. In such a setting, a sample can be selected that is statistically representative of the known characteristic, in other words, stratified along this dimension. For example, a researcher may posit a relationship between distance to market and degree of market participation. The sampling strategy would then stratify the sample along some measure of distance to market. Stratification comes in several forms and can be made proportionate to the size of the sampling unit or proportionate to variability within strata, as well as size (Miller and Salkind, 2002).

Clustered sampling typically relies upon geographically defined units from which households are selected, although clustering need not be undertaken along geographic lines. Clustering is often justified in order to keep costs at manageable levels. Nevertheless, it is important to keep the sample from becoming overly clustered, since an overly clustered sampling plan reduces the external validity or representativeness of the sample. See also Box 4.2, which presents the concept of two-dimensional sampling as a special case of stratified sampling.

Box 4.2 *Two-dimensional sampling*

Elements are typically identified by a number that represents a single position in a logical ordered sequence (for example, a household number). In some cases, however, the researcher may want to consider 'two-dimensional' populations. Such populations frequently arise naturally when the researcher has geographical or spatial concerns. For example, sample locations could be drawn from combinations of longitude and latitude.

Time may also be a key indentifying mark for the analyst, with each element in the population characterized by a time index in addition to a household identifier. In some cases, careful construction of a sampling procedure may allow the investigator to create an efficient survey design by drawing sub-samples of households at staggered intervals, thereby constructing a complete picture of activity across the year without burdening all respondents at each time step. Bethlehem (2009) outlines several methods for selecting a sample from a two-dimensional sample frame.

Staged sampling

Staged sampling corresponds to a procedure in which sampling occurs on different levels or in different waves, referred to as stages. For example, in a two-stage sample, the first stage requires that one selects a sample of clusters. These clusters are referred to as primary units and must be selected following some kind of formal sampling rule. Then, from each primary unit, the researcher selects a series of secondary units. These are the sample elements or observations. Again, a formal sampling rule is required for the second stage. Although multi-stage samples are possible, two-stage samples are most common. The most common implementation is when the primary unit consists of villages, and one randomly draws a small number of villages from the larger population of villages of interest. A small number of elements – for example, households – are then selected from within each primary unit. Alternatively, a small number of primary units can be combined with a large sample within each unit. The former approach typically provides for more precise estimators, since variance tends to be higher between villages than within villages, and is a common approach in social science fieldwork. One drawback of clustering is that the variance of estimators tends to be larger than in similarly sized random samples (Bethlehem, 2009). At the analysis stage, it may be necessary to account for this variance inflation by computing clustered standard errors, a relatively straightforward task in many statistical software packages.

Non-probability sampling methods

Convenience sampling

Convenience sampling is what Neuman (1991) refers to as 'haphazard' sampling. An example would be conducting interviews at a health clinic or church, or with individuals who volunteer to be interviewed. Samples selected in this way are inherently biased and for this reason are to be avoided wherever possible. However, cost considerations may lead a researcher to use a procedure that looks similar to convenience sampling, but is not. For example, in a location where households live in remote locations, local 'market days' may provide opportunities to interview in a short period of time a large number of respondents previously identified through some random selection process. Scarce enumerator resources can then be allocated to making household visits to those not interviewed at the market.

Snowball sampling

Snowball sampling is frequently employed when the target subjects for research are difficult to find, either because they represent a very small fraction of the underlying population or because they are largely hidden from view. The premise is to begin with an initial contact, either someone who matches the selection criterion or an individual who can serve as a 'key informant' to develop a list of potential respondents. Each individual who is interviewed is asked to provide the names of others who might serve as respondents. In this way, the sample grows like a rolling snowball, accumulating additional observations at each turn. The researcher terminates the process when a target sample size has been reached. Obviously, such sample designs are problematic because it is difficult to judge whether the sample is representative of the target population. Nevertheless, such an approach might be the only one available for sampling from some populations. For example, Shively et al (2010) used a snowball sampling approach to interview charcoal traders in Uganda, many of whom operate in a clandestine manner. Along similar lines, Boberg (2000) used a form of snowball sampling to study a wood fuel marketing chain in Tanzania. The survey started with producers, who identified the traders who had purchased their produce, who in turn identified the retailers to whom they sold the items. Despite the non-random nature of such samples, in some cases they may be the only way to collect relevant and useful information at reasonable cost.

Judgemental or quota sampling

Judgemental sampling relies on the judgement of experts in deciding who to interview. In most respects, such an approach is the polar opposite of probability-based sampling, since it relies completely on subjective opinion for choosing the sample. Proponents suggest that this approach reduces the likelihood of drawing a bad sample, but the pitfalls associated with this approach should be obvious, since it runs the risk of contaminating the sample with the bias of the experts. In some cases, however, it may be useful to use the judgement of experts when aiming to purposively include some particular aspect of the population. Quota sampling similarly follows a somewhat ad hoc approach, but is often guided by the goal of collecting information from a minimum number of respondents of a certain type or in a particular category.

Purposive sampling

Like judgemental sampling, purposive sampling is undertaken when probability sampling is either not available for some reason or is deemed unimportant for the research outcome. In some instances, researchers may actively look for a sub-group that appears to be somewhat typical of the population, or may purposively seek out a sub-group of the population for research focus, subsequently drawing a random sample from the purposively selected group.

How many: Sample size determination

In the case of non-probability sampling, especially when the goal of the research is qualitative rather than quantitative, a researcher may decide not to pre-specify a target sample size, but to instead continue collecting data, stopping only once

Box 4.3 *How many observations?*

If you plan to run regressions at the analysis stage, keep in mind that one's ability to generate point estimates of regression coefficients with statistical confidence is directly related to sample size. Larger samples produce larger t-statistics, by construction.

An informal rule of thumb used by many researchers is to ensure that $n/k > 15$, where n is the number of observations in a regression and k is the number of variables in the regression. In other words, each variable in a regression requires approximately 15 observations. With fewer observations than this, standard errors will be large and multicollinearity – the flip side of micronumerosity – will be a problem (see Goldberger, 1991). The upshot is that, to estimate a regression with 20 explanatory variables (not uncommon), you will probably need at least 300 observations. If your data set contains only 100 observations, do not plan on running regressions with more than six or seven explanatory variables.

his time or budget is exhausted or he feels confident that the data necessary to conduct the research investigation have been collected in sufficient quantity. More generally, however, determining the necessary and proper sample size will be an important research consideration in its own right, and sample size considerations will be based on concerns regarding the eventual statistical precision of parameter estimates derived from the sample (see also Box 4.3).

A number of methods have been developed for determining the proper size for a sample and, as a corollary, the statistical power of a sample for a given procedure. In some instances, simulation methods may be used *ex post* to build up a pseudo-population based on the sample, thereby providing one path by which to assess the power of one's sample. Power analyses, which are typically conducted prior to collecting data, seek to estimate the proper sample size based on acceptable probabilities of detecting effects or differences across groups. Statistical power is simply the probability of rejecting a null hypothesis when an alternative hypothesis is true. As Dattalo (2008, p16) argues, power analysis is compelling, since 'a study should only be conducted if it relies on a sample size that is large enough to provide an adequate and pre-specified probability of finding an effect if an effect exists'. Unfortunately, many individual researchers will be working with budgets that result in samples with relatively low power.

In general, the issue of sample size can be confronted either in a rigorous way, starting with assumptions about the population and using statistical methods to derive recommendations on sample size or, in a less formal way, using rules of thumb. One basic rule is that smaller target populations require larger sampling ratios than larger target populations. This is because there are rapidly diminishing returns to sample size with respect to accuracy. However, as Huck (2009) points out, the idea that larger populations always call for larger samples is a popular misconception. Neuman (1991) suggests a ratio of 30 per cent for small populations (those under 1000); 10 per cent for moderately large populations (those of, say, 10,000) and 1 per cent for large populations (those over 150,000). Collecting samples of 1000 or more observations, however, is likely to be beyond the budgets of most beginning researchers. Nevertheless, it is important to keep in mind that, for small samples, large gains in efficiency result from relatively small increases in sample size. Smaller samples can be justified when the underlying population is homogeneous. Conversely, when the population is quite diverse and/or when the researcher intends to study several variables simultaneously, as in regression models, relatively more observations will be required.

Approaching the issue formally, the United Nations (2008) presents a series of formulae that can be used to estimate a target sample size for purposes of collecting data on a population proportion. These formulae are based on collecting a sample that reaches a desired level of statistical precision. Unfortunately, what becomes immediately apparent when using such formulae is that the size of the target population relative to the total survey population plays a crucial role in the choice of a sample size. In general, if the target population of interest is a relatively small proportion of the total survey population, then the required sample will be relatively large. Conversely, if the target population of interest is a relatively large proportion of the total survey population, then the required sample will be comparatively smaller. As an example, standard calculations suggest that if the target population of agricultural households were believed to be 90 per cent of rural households (the survey population), then the appropriate sample size to reach a 95 confidence level for sample statistics would be approximately 300. On the other hand, if the sample were being drawn from rural households, but agricultural household made up just 10 per cent of the rural population, the appropriate sample size would expand to more than 2000! The general principle illustrated, therefore, is that much larger samples are required when the population of interest constitutes a small proportion of the sample population. This is one reason why it is often difficult to generalize from small samples to large populations. Another factor that influences the choice of sample size is the anticipated rate of non-response, since one must plan for some losses in observations during the survey.

Some practical concerns

Village norms and random sampling

When working in rural villages, the concept of random sampling may not be easily understood by participants, including village leaders. Notables in the village may feel offended at not being included. Others may suspect that households have been selected based on favouritism (or for other more mysterious reasons). Thus it may prove useful to make selection transparent by inviting villagers to be present when samples are drawn. Those selected via lottery may feel lucky and participate with enthusiasm.

In some cases, it may be important to include the local leaders, such as the headman or village chief to ensure his cooperation and also to facilitate cooperation from those selected. For this reason, even if the household of the village chief is not among the households selected at random for the sample, one may still want to include that household in the interviews and state in public that his household is included. However, if included in this way for purposes of facilitating work in the village, that household should be dropped from the analysis.

Refusal, non-response and attrition

In most surveys, some of the selected households may refuse to participate for various reasons. Thus when doing the random selection, one should always have a few extra households (perhaps 5–10 per cent) available (and on reserve) who have been selected using the same procedures as those in the sample. These can be used to replace any households that refuse to participate. However, one should always attempt to minimize refusals and drop-outs by approaching households in an appropriate manner, carefully explaining the purpose of the survey and encouraging participation. If a large proportion of selected households refuse to participate and must then be replaced by those willing to participate, this can cause serious problems related to selection bias, even if replacements are chosen at random (see Huck, 2009, for an extended discussion of this point). The extent of such bias will depend on the factors that distinguish participants from non-participants, for example, people involved in illegal forest extraction may be more inclined to refuse to participate. Since no information is usually collected from those who refuse to participate, the sources, direction and magnitude of biases cannot be determined.

A related problem is those who drop out of a multi-round survey after the survey has started. This is called attrition. Some safeguards to attrition problems are to begin with an overall sample that is larger than the target sample. However, as with those who initially refuse to participate, high drop-out rates have the potential to generate selection bias. The general suggestion is, of course, to try to minimize attrition (the number of drop-outs). Nevertheless, most researchers implementing multi-round surveys must accept some degree of attrition: respondents lose interest, individuals die and people move. Methods for testing for and adjusting for attrition are available, but they are somewhat complex and beyond the scope of this chapter. Some guidance and empirical evidence is provided by Alderman et al (2000). Fortunately, evidence suggests that in many contexts, attrition may not present as many difficulties for statistical inference as is sometimes feared.

Measuring impacts

One important issue to keep in mind when developing a sampling strategy is that the extent to which a researcher can answer specific causal questions about the impacts of external forces on a population will be sharply delimited by her ability to formulate alternative hypotheses regarding cause and effect, and to isolate specific explanatory factors for investigation. Questions regarding impact may centre on policies. For example, one might be interested in the impact of forest policy reforms on forest use, or may want to study the impact of agricultural innovations on forest clearing by farmers. In the world of programme evaluation, one conducts an impact evaluation study to assess the degree to which changes in observed outcomes can be attributed to changes linked to the programme, rather than to other factors that might affect outcomes. Impact evaluations seek to answer the question, 'does a project, programme or intervention work better than nothing at all?' As Ferraro (2009) suggests, an answer to such a question requires that the researcher know

Box 4.4 *Experimental treatments and natural experiments*

Increasingly, researchers are attempting to measure the impacts of projects by randomizing project participation (in advance) and then observing changes in outcomes (afterwards). The attraction of this approach, called a randomized controlled trial (RCT) stems from the fact that, if a researcher can carefully control how an experiment is designed and how data are collected, then variation can be introduced through random assignment of individuals to treatment and control groups (and through that channel alone). However, experiments can be costly to administer, are only practical when a small number of treatment variations are being considered and raise a number of important ethical issues.

External events that are truly exogenous in their origin can be likened to experimental treatments. In rare instances researchers may encounter a *natural experiment*, in which some force is encountered that originates outside the control of the research subjects or the researcher, but nevertheless separates otherwise similar populations into treatment and control categories. These outside forces may result from natural occurrences, economic or political events, or accident. For example, a bridge may wash out during a storm, leaving one community unaffected but an adjacent community temporarily isolated, or a new law may be enacted in one jurisdiction and not in an otherwise similar area. In most natural experiment cases, a researcher must have specific knowledge about the groups under investigation before and after the event. As an example, Shively (2001) studied the introduction of irrigation in the Philippines as a natural experiment to measure the impact of irrigation on labour market participation and forest clearing. Despite the potential usefulness of controlled and uncontrolled (or natural) quasi-experiments, however, caution must be exercised since, as with all similar types of 'before versus after' or 'with versus without' investigations, the validity of one's statistical inferences rests on the specific assumption that the assignment of households or individuals to treatment and control groups is not related to other factors that may have determined outcomes (Ferraro, 2009).

something about what outcomes would have looked like in the absence of the intervention. If the research goal is to study the impact of programme participation or a policy change, it is extremely important when designing the sampling strategy that the survey seeks to isolate external effects by collecting data on cotemporaneous factors that are believed to be correlated with external interventions and observed outcomes. It is also essential to know what features of the population may have led to selection bias. *Selection bias* can occur when individuals voluntarily choose (in other words, self-select) whether to participate in a programme, and the characteristics of those who choose to participate differ systematically from those who choose not to (for some approaches to measuring impacts, see Box 4.4).

Documentation

As research proceeds, and upon completion, it is important to clearly document the sampling methods used at the data collection stage. Among the questions the sampling documentation should seek to answer are the following:

- What was the target population for the survey?
- How was the sample frame chosen?
- Was there overcoverage or undercoverage of the population?
- When and where were the data collected?
- If stages were used, how was selection done at each stage?
- What were initial and final sample sizes? What explains any discrepancy?
- What are the known biases in the data?
- What problems, if any, were there with attrition or non-response?
- Of what population, time and place are the data likely representative?

Providing answers to these questions communicates clearly to colleagues, potential reviewers and those who might later use the data the exact strengths and weaknesses of your data.

Conclusions

This chapter has tried to provide some formal guidance on the issue of sampling. Any sampling strategy should be aligned to the type of research question (or questions) and hypotheses to be investigated and the kinds of analyses foreseen. While it is generally true that larger samples provide greater opportunities to engage in more rigorous numerical analysis, statistical sophistication alone is not sufficient justification for collecting more data. Many useful and important studies have been undertaken with surveys that were relatively modest in size, but carefully conducted. An important issue in deciding on a sampling strategy is variation. Research often focuses on explaining differences in behaviour and outcomes by way of differences in material conditions and other framing characteristics. If the dependent variable has no variation, there will be no variation to explain. Similarly, explanatory variables that do not vary across the sample cannot be expected to provide much explanatory power. Finally, this chapter has argued that, while non-probability sampling methods may be useful in specific settings, they should be approached with caution and employed only when random sampling within sub-samples cannot be used or justified. Probability sampling is the default option, unless there are strong reasons to proceed otherwise.

Key messages

- Be careful to consider the research questions and hypotheses and target population when deciding on the sampling strategy.
- Make sure that the sampling strategy facilitates the needed variation in data.
- Choose probability sampling unless there are compelling reasons to do otherwise.

References

Alderman, H., Behrman, J. R., Kohler, H.-P., Maluccio, J. A. and Watkins, S. C. (2000) 'Attrition in longitudinal household survey data, some tests for three developing-country samples', Policy Research Working Paper No 2447, World Bank, Washington, DC

Bethlehem, J. (2009) *Applied Survey Methods: A Statistical Perspective*, John Wiley & Sons, Hoboken, NJ

Boberg, J. (2000) *Woodfuel Markets in Developing Countries: A Case Study of Tanzania*, Ashgate, Aldershot

Byrne, D. (2002) *Interpreting Quantitative Data*, Sage Publications, London

Cochran, W. G. (1977) *Sampling Techniques*, Wiley, New York

Czaja, R. and Blair, J. (2005) *Designing Surveys: A Guide to Decisions and Procedures*, Pine Forge Press, Thousand Oaks, CA

Dattalo, P. (2008) *Determining Sample Size: Balancing Power, Precision, and Practicality*, Oxford University Press, Oxford

Ferraro, P. J. (2009) 'Counterfactual thinking and impact evaluation in environmental policy', in Birnbaum, M. and Mickwitz, P. (eds) *Environmental Program and Policy*

Evaluation: Addressing Methodological Challenges, New Directions for Evaluation, vol 122, pp75–84

Goldberger, A. (1991) *A Course in Econometrics*, Harvard University Press, Cambridge

Gregoire, T. G. and Valentine, H. T. (2008) *Sampling Strategies for Natural Resources and the Environment*, Chapman & Hall/CRC, Boca Raton, FL

Hoyle, R. H. (1999) *Statistical Strategies for Small Sample Research*, Sage, Thousand Oaks, CA

Huck, S. W. (2009) *Statistical Misconceptions*, Routledge, New York

Kline, R. B. (2009) *Becoming a Behavioral Science Researcher: A Guide to Producing Research that Matters*, The Guilford Press, New York

Merkens, H. (2004) 'Selection procedures, sampling, case construction', in Flick, U., von Kardorff, E. and Steinke, I. (eds) *A Companion to Qualitative Research*, Sage, London, pp164–171

Miller, D. C. and Salkind, N. J. (2002) *Handbook of Research Design and Social Measurement*, sixth edition, Sage, Thousand Oaks, CA

Neuman, W. L. (1991) *Social Research Methods, Qualitative and Quantitative Approaches*, Allyn and Bacon, Boston, MA

Särndal, C.-E. and Lundström, S. (2005) *Estimation in Surveys with Nonresponse*, John Wiley & Sons, Chichester

Sen, R. S. and Sharma, N. (2008) 'Unravelling the mystery of creativity', in Anandalakshmy, S., Chaudhary, N. and Sharma, N. (eds) *Researching Families and Children: Culturally Appropriate Methods*, Sage, Los Angeles, CA, pp165–191

Shively, G. E. (2001) 'Agricultural change, rural labor markets, and forest clearing: An illustrative case from the Philippines', *Land Economics*, vol 77, no 2, pp268–284

Shively, G., Jagger, P., Sserunkuuma, D., Arinaitwe, A. and Chibwana, C. (2010) 'Profits and margins along Uganda's charcoal value chain', *International Forestry Review*, vol 12, no 3, pp271–284

Tuli, M. and Chaudhary, N. (2008) 'Cultural networks, social research and contact sampling', in Anandalakshmy, S., Chaudhary, N. and Sharma, N. (eds) *Researching Families and Children: Culturally Appropriate Methods*, Sage, Los Angeles, CA, pp53–65

United Nations (2008) 'Designing household survey samples: Practical guidelines', Studies in Methods Series F, No 98, Department of Economic and Social Affairs, Statistics Division, United Nations, New York

Wysocki, D. K. (2008) *Readings in Social Research Methods*, Thomson Wadsworth, Belmont, CA

Box 4.5 *Further reading*

The classic textbook on sample selection is Cochran (1977). A more recent treatment is provided by Czaja and Blair (2005). Bethlehem (2009) presents a comprehensive introduction to applied survey methods and also provides a fascinating review of the history of survey research and an extensive treatment of complex sample designs and population estimators. Neuman (1991) offers a succinct introduction to sampling within the overall context of applied social research methods. A 2008 United Nations handbook covers a number of practical concerns, especially with regard to large surveys (but with equal relevance to small surveys). A somewhat sceptical (but refreshing) view of probability-based sampling is provided by Byrne (2002). Merkens (2004) discusses sampling for case studies. Readers who are considering setting up a formal experimental or non-experimental design are encouraged to consult Kline (2009), who outlines a number of design and analysis considerations for such applications. Wysocki (2008) provides a useful set of readings related to non-probability sampling and offers examples that highlight many of these approaches to data collection. Dattalo (2008) provides a comprehensive review of sampling issues, including an extended discussion of computer-based methods for conducting sampling diagnostics. He also provides a number of worked numerical examples. Särndal and Lundström (2005) outline a number of strategies available for dealing with problems of non-response. Finally, for researchers who require an understanding of resource (as opposed to human) sampling for ecological, forestry or environmental analysis, a very comprehensive treatment of that topic is provided by Gregoire and Valentine (2008).

Chapter 5

Collecting Contextual Information

Georgina Cundill, Sheona Shackleton and Helle Overgaard Larsen

---

A good decision is based on knowledge and not on numbers.
Plato (428–348 BC)

Introduction: The importance of understanding context

Quantitative measurements of livelihoods and environmental dependence among rural households only really gain full meaning when interpreted within the broad situational context from which they emerged and of which they are a part. Livelihoods are shaped by a multitude of forces and factors that are themselves constantly changing (DFID, 2000). Communities and user groups dependent on rural natural resources operate within a context that is broadly defined by biophysical, demographic, cultural, technological, political and market-related factors; by the nature of state agencies, policies and strategies; by the legal and institutional setting; by historical processes; and by the involvement of various external actors and groups (non-governmental organizations (NGOs) and governments) in the development landscape (Agrawal and Angelsen, 2009). Contextual factors, as well as internal household dynamics, shape how different households construct their livelihoods and utilize environmental resources. Resource endowment and market factors, for example, can play a major role in determining the importance of natural resources to households and the levels of exploitation and management of the resource base. This is illustrated in the following example.

Mbuti Pygmies living in the Central African Ituri forest follow two different hunting strategies. Bands in the north-eastern part of the forest hunt with bows and arrows, while bands in the south-western part have adopted the technology of net hunting. Both types of bands acquire crop foods through trade with neighbouring agriculturalists, but the bands in the south-western part do so for a longer period of the year. What may be the reasons that some Pygmy bands retain their traditional hunting practice while others abandon it? Milton (1985) argues that lower availability of wild food and wildlife in the south-western part of the forest have made the Pygmy bands living there more motivated to secure a stable high-quality food source through trade of surplus meat for agricultural products. This explanation is supported by evidence regarding the quality of soils, the number of associated tree species and the level of secondary compounds such as tannins in the foliage of trees and other plants.

A survey of the livelihood activities of the Mbuti Pygmies would yield information on the number of days spent hunting, numbers and types of animals caught and the amount of meat traded for agricultural products. One of the strengths of the survey approach is arguably that a large sample of people's reported behaviour is recorded and real-life potentials and challenges to livelihoods can therefore be identified at relatively low costs as compared to using, for example, participant observation. A potential limitation of quantitative livelihood surveys is, however, that without an understanding of the context in which livelihoods are situated, data interpretation is speculative. If one seeks to understand the options available to the Mbuti Pygmy hunters, numbers on hunting days and catches are not enough – they simply establish the fact that one group hunts more than the other with no hints as to the reason for this.

A second limitation to quantitative approaches is that they will not show, with any great detail, how people's choices are embedded in culture and history, as well as within institutional, economic and political settings at multiple scales (Ellis, 2000; Murray, 2001). Thus, these approaches have been criticized for failing to consider issues of power and broader social relationships (Moser et al, 2001; Conway et al, 2002). A third potential limitation of the quantitative household survey is one already alluded to above: the static nature of the data, as most surveys are conducted only once. A common use of survey data is to make evidence-based forecasts. Such predictions are critical parts of informing policy and they require a good understanding of trends over time – and particularly how important drivers of change – for example, policies – affect livelihoods. As this is rarely captured by one-off quantitative surveys, it is critical to incorporate information at coarser temporal and spatial scales when interpreting local data sets and to gather data on the nature of drivers of change.

Contextual information is important at several stages of the research process. At the outset, contextual information is key to developing quantitative research instruments that are able to collect relevant information. For example, contextual information may serve to improve the quality of survey data when knowledge on typical livelihood activities is used in questionnaire design (see Chapter 7). Furthermore, contextual information can be highly important to the actual implementation of the questionnaire survey through sensitizing the research team to local cultural and political issues that may affect the research (see Chapters 10 and 11). Finally, once data has been collected, contextual data is important for interpreting findings and placing these into perspective. While regression analyses establish correlations, contextual information often proves useful when trying to establish causal relations behind the observed correlations. Furthermore, contextual information may complement the survey data in the following ways: by filling gaps in the information based on the survey; for comparing findings across sites; for generalizing and scaling-up findings; and for making future predictions. Examples of contextual information that is usually not covered by a household survey include: local notions of wealth; wealth dynamics; and in-depth understanding of strategies for coping with risks and shocks.

Researchers are often intimidated by the huge variety of contextual variables that are relevant to a given study and by the task of selecting the correct scale at which to start. For example, if a community reports declining prices for agricultural produce, leading to abandonment of fields and increased migration to cities, would you start by looking at changes in local markets, regional markets, or should you be looking at national agricultural policies? A critical question, then, is what to measure and where to start. The Sustainable Livelihoods Framework (SLF) (DFID, 2000) can provide a useful guide when navigating the often complex interactions that are observed. It is premised on the understanding that rural livelihoods are diverse, complex, dynamic and socially differentiated, and that people's access to different forms of livelihood capital and, consequently, the types of livelihood strategies and activities they engage in are mediated and influenced by the economic, political, ecological and institutional environment within which they find themselves. In this chapter, we use the SLF to describe contextual data in relation to several aspects of community life that may inform and complement quantitative household surveys. The types of contextual information required, the steps and methods for gathering this and the way in which it can be used to complement and qualify household survey data are discussed. Data collection methods are briefly introduced and references to relevant sources provided.

A framework for gathering contextual information

The SLF highlights household productive activities in relation to a vulnerability context, as well as the policies, institutions and processes that mediate access to different sources of livelihood capital (Hulme and Shepard, 2003, Figure 5.1). Five types of capital are identified: natural, human, physical, social and financial (Figure 5.1, definitions in Box 5.1). Gathering such information is key to

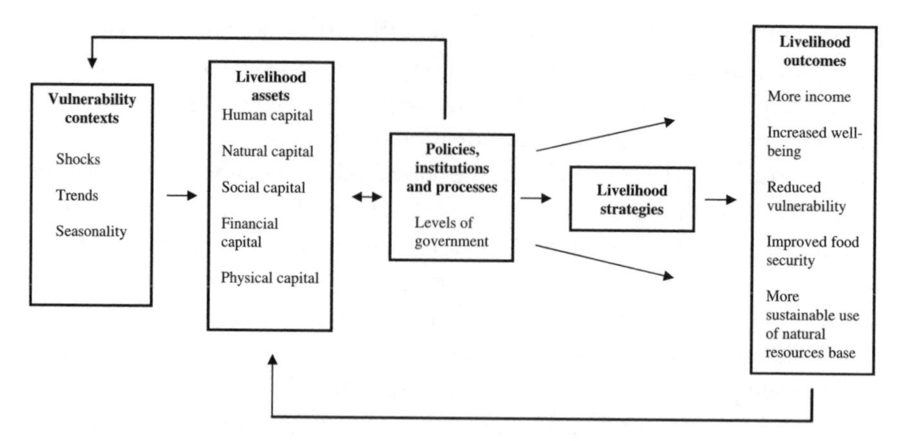

Figure 5.1 *The Sustainable Livelihoods Framework (SLF)*
The arrows within the framework are used to indicate a variety of different types of relationships, all of which are highly dynamic. None of the arrows imply direct causality, though all imply a certain level of influence.
Source: Adapted from DFID (2000)

understanding differential access to and use of assets, and thus different portfolios of livelihood strategies amongst surveyed households.

Box 5.1 *The five capitals of the Sustainable Livelihoods Framework (SLF)*

Human capital: The skills, knowledge, ability to labour and good health that together enable people to pursue different livelihood strategies and achieve their livelihood objectives.

Social capital: The social resources upon which people draw in pursuit of their livelihood objectives, including social networks, membership of formalized groups and relationships of trust, reciprocity and kinship.

Natural capital: The natural resource stocks from which resource flows and services useful for livelihoods are derived. Examples of resource stocks and the services derived from them include: land, forests, marine/wild resources, water, air quality, erosion protection, storm protection and biodiversity, among others.

Physical capital: The basic infrastructure and producer goods (for example, equipment, roads, transport, communications) needed to support livelihoods.

Financial capital: The financial resources that people use to achieve their livelihood objectives (for example, savings, credit, cash income).

Source: DFID (2000)

While quantitative household surveys can provide some data on financial, physical and human capital at the household level, it may be desirable to collect in-depth information regarding natural and social capital, or information about the availability of different forms of capital at the scale of community and district. Obtaining information related to, for example, the abundance and state of natural resources, access to infrastructure and services, local organizations, social networks and community cohesion would be critical, but difficult to gather using quantitative surveys. In the next section we shall describe the important contextual information that should ideally be obtained for each of these five forms of capital and the methods for collecting it.

The framework also recognizes the importance of the 'vulnerability context' within which households and communities operate, and how this may influence multiple dimensions of people's livelihoods. The vulnerability context can include intermittent and unpredictable hazards such as droughts and floods, or more long-term stresses and risks such as ongoing health problems (malaria, HIV/Aids and other tropical diseases), poverty, lack of development, conflict and war, or trends such as climate change, land and resource degradation, population growth and pressure, and so on. Understanding the vulnerability context is key to understanding adaptive capabilities, coping strategies and the role of environmental resources in securing livelihoods and potentially providing a safety net.

The concept of context in the SLF is meant to include many factors not explicitly mentioned and often forgotten (Scoones, 2009). Relevant to our purpose are local knowledge, institutions understood as shared meanings, norms and behaviour, landscape changes, government service delivery and technological innovations that may influence how people use their resources.

In addition to the factors mentioned above, understanding historical trends is fundamental to understanding possible future trajectories and change. Understanding these possible trajectories is, in turn, key to converting research outcomes into tangible recommendations for policy and practice. We have sought to build an understanding of historical processes and change into all the spheres of contextual information discussed below.

Collecting contextual information

Contextual information can be obtained from many primary and secondary sources, and usually a multi-source approach is required. At a national level, policies, acts and regulations, and government strategies can provide important background information. At a more local level maps, census statistics, previous research reports, NGO reports and other secondary data can be invaluable in gaining a better understanding of an area and its history. Primary data may also be collected at community level using participatory techniques (Participatory Learning Approaches (PLA), Participatory Rural Appraisal (PRA)), observation, key informant interviews with people within and outside the community (community leaders, extension officers, NGO representatives, middlemen), and focus group discussions with specific groups of interest (women, specific resource users). A key issue when deciding what contextual data collection method to use is at what level variation in the data is needed. For example, when studying household forest incomes it will be relevant to investigate whether access to forest resources differs among ethnic or wealth-based groups. In that case, focus group discussions are more suitable than community level inquiry. In this chapter we focus on the collection of primary local level contextual information, with some reference to secondary sources. We have organized our discussion according to the 'livelihood assets', 'policies, institutions and processes' and 'vulnerability contexts' components of the SLF (Figure 5.1), as it is these that influence livelihood strategies and outcomes. We begin by discussing the ways in which researchers can collect contextual information regarding the various forms of capital.

Learning about natural capital

When interpreting household survey data regarding natural resource use and resulting livelihoods outcomes, it is important to consider what natural resources people actually have access to and what influences their decisions regarding what to use. For example, decisions may be based on availability, preferences and/or accessibility, where the latter can differ among individual households due to rules, customs or other factors. Resource endowments refer to the abundance and availability of natural resources in a given locality, and a variety of participatory and non-participatory methods can be used to gather information about these endowments.

Maps, satellite images and aerial photos: An initial overview of a community's biophysical setting and natural resources can be achieved by looking at maps, satellite images and aerial photographs obtainable from various sources such as local libraries and agricultural departments or bought or downloaded for free from the internet. Locally specific information can be gathered using participatory approaches by drawing maps with key informants identifying the location of various features of interest. For example, different types of agricultural fields, different forest types, grazing areas, different ecosystems or different areas from which natural resources are extracted. Basing participatory resource maps on aerial photographs yields geographically precise information and allows for calibrated perceptions of observed features, such as forest degradation levels. In Nepal, photo interpretation was observed to include rather than exclude people without formal education, including women, when delimiting Nepalese community forest boundaries and discussing the state of the forest resource (Mather, 2000). In addition to the mapping exercises, transect walks and resource inventories can highlight the location of resources, their abundance, their distance from homesteads or settlements and their accessibility, as well as providing a means for checking the accuracy of the maps.

Ranking exercises: Understanding local preferences for particular natural resources, for example, particular species of fuel wood, is an important contextual issue that provides insight into the values, often intangible, that influence the use of natural resources observed during household surveys. Ranking exercises are one of the most effective methods of understanding preferences and are generally conducted during participatory workshops or focus group discussions. Ranking involves placing issues or objects in order of significance (Theis and Grady, 1991). Ranking is not only useful for understanding natural resources, it can equally be used to determine the relative importance attached to activities, environmental attributes and stakeholders (see, for example, Richards et al, 1999; DFID, 2000). The actual rank is not as important as the comments and the debates that are generated by the exercise (Sithole, 2002). Matrices can also be used, and provide a useful means of summarizing a great deal of information (Nori et al, 1999). Matrices may be either simple checklists or complex tables with rows and columns containing information regarding, for example, people's preferences and reasons for these preferences (Theis and Grady, 1991; Rowley, 1999; Sithole, 2002; Box 5.2).

Trend lines: The above contextual information is useful, but it is not sufficient without an understanding of the associated trends. A temporal dimension can be added to any of the exercises referred to above by conducting trend line exercises or historical mapping (Chambers, 1992). Historical mapping provides visual evidence of changes that have occurred over time, and in this way helps to identify causes of current environmental change (Borrini-Feyerabend, 1997). Trend lines can be used to illustrate changes in availability of species, for example, and the reasons behind these changes. Trend lines are sets of key dates in chronological order with a score representing relative changes in key variables; in the case of natural capital, these could include water quality, forest health and species availability (see Figure 5.2). Trend lines can also be used to discuss issues varying from access to markets and credit to changes in land use, customs and practices, population, fuels used, and so on (Chambers, 1992). This information serves to guide the researcher in the search for additional contextual information,

Box 5.2 *Example of a matrix exercise summarizing what crops are cultivated, the most important reason for their cultivation and the most important markets for these crops*

Key: ✓crop is grown for this purpose
--crop is not grown for this purpose
✓✓ the most important reason for cultivation/most important market

Crops	Household use	For sale	Sold in town	Sold in village
Maize	✓	✓✓	✓	✓✓
Potatoes	✓	✓✓	✓✓	✓
Pumpkin	✓	✓✓	✓✓	✓
Beans	✓	✓✓	✓✓	✓
Green beans	✓✓	✓	--	✓✓
Melon	✓✓	✓	--	✓✓
Wheat	✓	✓✓	--	✓✓
Wild spinach	✓	✓✓	✓	✓✓

Source: Cundill (2005)

because discussion will inevitably be regarding what caused the changes in natural resources, and this often relates to social factors. Agricultural census data is a key source of historical data and can be used to verify information gathered during trend line exercises or other participatory exercises.

Historical mapping: involves drawing maps of the same area for different dates in the past to illustrate key changes. For example, changes over time in the area and location of arable fields, hunting areas, forest cover or grazing land can be illustrated in this way. Having drawn a map or identified features on historical aerial photographs, these maps can be used as a basis for further background data collection through key informant interviews. These resource dynamics may indicate changes in livelihood options as the basis for some strategies disappearing and new ones emerging (Ellis, 2000).

Figure 5.2 *A trend line exercise showing declining water quality in a village in the Eastern Cape, South Africa between 1960 and 2000*
Source: Christo Fabricius

Learning about physical capital

The infrastructure available to people influences their ability to pursue particular livelihood strategies. For example, access to agricultural machinery or ploughing oxen will influence the ability of some households to cultivate fields. The presence of roads and/or navigable rivers will determine market access and therefore market-based activities, while the existence of dams and/or irrigation systems will influence what is planted as well as risk and vulnerability to drought. It is important to note that access to communally owned or managed infrastructure, for example irrigation, may be differentiated through, for example, elite capture (see Balooni et al, 2010) or as the location of land plots relative to the irrigation outlet influence the benefits derived (Makurira et al, 2007).

While individual household assets can be listed during surveys, access to communal and regional infrastructure is also important. Topographical maps are a good starting point to identify settlements, dams and road networks. Aerial photographs can be used to identify market facilities. More detailed information can be gathered during participatory mapping exercises, key informant interviews with the members of agricultural cooperatives or committees and also simply by looking around. Clinics, schools, community halls, water taps and fences are all good examples of physical capital that will influence livelihood options and the use of natural resources. Participatory mapping of neighbourhoods can be useful to plan sampling procedures before surveys begin.

Learning about social capital

Social capital includes both tangible relationships such as membership of formal and informal groups, and intangible relationships such as kinship, trust and reciprocity networks. Understanding these relationships helps researchers to make sense of household vulnerability because, in many instances, these networks enable people to cope with shocks and change.

Some basic information regarding kinship, trust and reciprocity can be collected using standard survey techniques by, for example, including questions about reliance on family and friends during crises (see also Chapter 7). More in-depth and semi-structured interviews are, however, necessary to come to terms with the ways in which these networks are made use of, or how exclusion from them affects resource use and reliance patterns. For example, in Pondoland in the Eastern Cape, South Africa, Kepe (2003) and Makhado and Kepe (2006) found that women's participation in weaving activities, the volume of products produced and resulting incomes earned from crafting were closely related to 'ownership' of wetland reed (*Cyperus textilis*) beds. The information about the ownership was not yielded in the formal household survey but identified through PRA exercises and in-depth interviews with a selected group of women. In this region, unlike other parts of South Africa, reed beds along rivers and streams within the village commons are 'quasi-privatized' and under the stewardship of female members of households. These women lay claim to the beds by planting *Cyperus textilis* rhizomes. This planting provides tenure rights (including inheritance) over the planted area as long as it is maintained; as soon as a site is neglected or abandoned, a new owner can take possession. Women without reed beds gain access to reeds via the informal networks they are part of. Kin and close friends often harvest for 'free', while in other situations a 50–50 share cropping system may apply, where labour is provided as payment for reeds harvested or the harvester may weave a mat for the reed bed owner in exchange for access. Sometimes women living close to the streams and rivers may look after a bed for the owner, who may reside a long distance away, and thus gain permission to harvest.

Information regarding tangible networks, such as local leadership, community groups and links to local government or NGOs, can be gathered using key informant interviews. Key informants to focus on would usually include committee members and other community leaders. For information regarding village governance, additional key informants who are not members of official organs should be consulted for verification. In participatory workshops, Venn diagrams are a powerful tool used to diagrammatically represent relationships between social groups (Box 5.3; Asia Forest Network, 2002; DFID, 2000) and can be used to explore relationships between different actors and organizations.

Learning about human capital

Information regarding human capital is normally included in household surveys in terms of education levels in the household, any formal or informal training received by household members and skills held by household members such as weaving and carving. Information regarding household health and the ability of household members to work is an important aspect of human capital and should be included in household surveys. Contextual information supporting, *ex ante*, the questionnaire design could include what types of skills and training are demanded in a particular community and what skills are in low supply. Key informant interviews with local leaders, extension agents and NGO staff will provide information on any training and skills development activities in the area, as well as the number of primary and secondary schools that children have access to. Useful secondary data sources, such as local census data, will enable you to put this information into the broader context by providing information on education and literacy levels and common types of formal employment.

Box 5.3 *Example of Venn diagram illustrating engagement with forest utilization*

A Venn diagram can be used to map relations between actors, such as groups of natural resource users. When constructing the diagram as a participatory exercise with resource users, both the resulting diagram and the discussions entailed are useful sources of information.

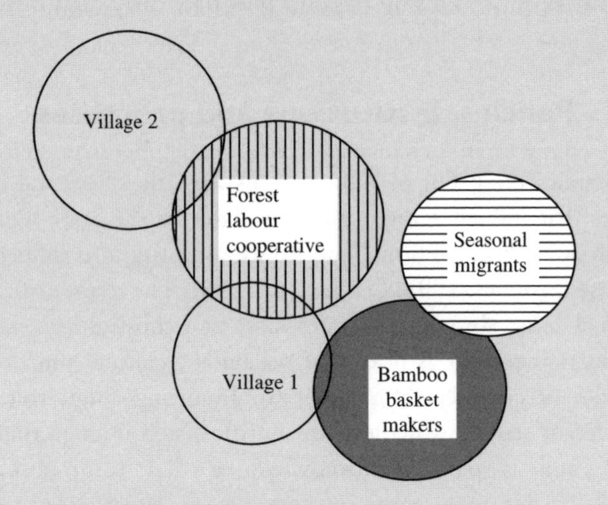

Figure 5.3 *Example of a Venn diagram*

The Venn diagram consists of circles representing the various actors involved in or influencing resource use that are drawn on a piece of paper or on the ground. It may be useful to reflect the spatial positions of users when positioning circles. The size of circles typically reflects the importance or size of the actor, overlaps represent shared membership. In this example, inhabitants from two villages are members of a forest labour cooperative while seasonal migrants and inhabitants from village 1 participate in bamboo basket-making.

Source: Asia Forest Network (2002)

General information on the health status of community members can be obtained from local clinics and women's groups.

Learning about financial capital

In conjunction with household surveys documenting income and expenditure at the household level, census data can be a key source of information regarding financial capital, particularly in terms of employment rates and income levels in the broader area. Key informant interviews with local government officials, community leaders and NGOs active in the area can provide an idea of the level of external financial input and support taking place through development projects. Local key informant information can provide an understanding of the availability of loans. In the absence of formal loans from banks and other credit institutions, villagers have to rely on informal loans but their availability may depend on, for example, kinship ties and they may carry high interest.

Policies, institutions and processes

Livelihood studies often fail to fully acknowledge the structural barriers and opportunities that influence the options available to the poor (Du Toit et al, 2007; Gough et al, 2007). Formal and informal institutions influence patterns of resource use (Scoones, 1998; Leach et al, 1999). The term 'institution' refers to both formal constraints such as rules, laws and constitutions, and informal rules and constraints such as norms of behaviour, cultural practices and self-imposed codes of conduct (North, 1990). Inquiring about rules governing access to different resources is therefore a critical step in collecting contextual information, and is part of acknowledging that communities are not homogenous entities with common interests or equal access to resources (Agrawal and Gibson, 1999; Leach et al, 1999; Ribot and Peluso, 2003).

A useful starting point in elucidating differing levels of decision-making power, existing organizations and institutions, and how people's choices are constrained or enabled by these structures is to conduct a Venn diagram exercise with a small group of actors (see Box 5.3). This exercise can be repeated several times with different user groups to get a more nuanced idea of power and rules governing resource use. These exercises can be accompanied by key informant interviews with natural resource users, customary and elected leadership operating at various levels, for example, traditional or elected leaders at the community level and local forestry officials, elected councillors and municipal managers at local government level.

The policy and legislative environment can also have a major impact on what happens at the local level and can influence who has access to what forms of livelihood capital. Policies that support decentralization and devolution can, for example, both increase and decrease local people's access to natural resources (Shackleton et al, 2002). Understanding access to and the security of tenure that a household has over the land that they cultivate or the resources that they use is often critical in understanding the investments made by a household in developing that resource and the benefits they obtain from it. Desktop studies involving consultation of national, provincial and municipal policies, legislation and strategies will usually be useful to understand *de jure* (legal) access and tenure arrangements. This should, however, be combined with key informant interviews with users and leadership (customary and official) to understand the overlaps (if any) with de facto tenure in a given setting. Often informal tenure arrangements may emerge at a local level such as described earlier in the Pondoland *Cyperus textilus* example. In-depth interviews and group discussions with resource users can also be useful to elucidate the level of security that resource users feel about their access to resources and any conflicts or struggles around this access. Lund and Treue (2008), for example, combine a household survey and in-depth interviews to show how a group of villagers that are highly dependent on charcoal production faces strict enforcement of rules due to a long-standing conflict with the village leadership.

Trends, shocks and vulnerability

Livelihoods are affected by trends as well as by shocks and seasonality over which people often have limited or no control. In many instances, the livelihood data recorded during surveys will reflect either short-term responses to shocks and/or seasonality, or long-term adaptive responses to trends such as population growth or technological innovation. Understanding the ways in which these factors influence livelihoods, and local responses to them, is critical to understanding data from quantitative surveys that provide snapshots of households' livelihoods. Box 5.4 provides a limited list of examples of these factors.

Secondary data sources can provide useful information relating to trends and shocks. Trends in population, education, employment and health can often be obtained using local census data. The local weather station should have long-term rainfall and temperature data, while clinics may have long-term data regarding health, and agricultural offices may have information regarding trends in production and prices for produce.

To obtain primary data, many of the methods discussed previously will be useful. Key methods include trend line exercises and historical mapping. The vast majority of the information relevant to the vulnerability context will emerge from the debates and discussion between participants involved in participatory workshops dealing with other parts of the livelihoods framework, for example, when talking about natural capital and changes in either abundance or availability. Seasonal calendars are another useful method that can be used to begin discussions regarding how livelihoods shift during different times of the year. Seasonal calendars are a series of different diagrams showing the main activities, problems, and opportunities throughout the annual cycle, and are most useful in identifying the months or periods of greatest vulnerability or difficulty (Theis and Grady,

Box 5.4 *The vulnerability context*

Trends

- Population trends.
- Resource trends (including conflict).
- National/international economic trends.
- Trends in governance (including politics).
- Technological trends.

Shocks

- Human health shocks.
- Natural shocks.
- Economic shocks.
- Conflict.
- Crop/livestock health shocks.

Seasonality

- Of prices.
- Of production.
- Of health.
- Of employment opportunities.
- Of natural resources availability.

Source: DFID (2000)

1991). Calendars can be based either on seasons or on months of the year and are useful in identifying the temporal distribution of rainfall, water requirements, crops, labour, fuel, migration, income, and so on (Chambers, 1992).

Conclusions

Contextual information is important at various stages of the research process, from designing household surveys and selecting the research site to interpreting, understanding and drawing conclusions on the basis of quantitative survey data. The SLF is a useful guide that can serve as a compass to orientate you in the collection of contextual information and help you better understand the factors that influence the livelihood choices, strategies and outcomes (Box 5.5) of different households. Conducting as many interviews yourself as possible, listening to people's stories about their lives and following up on issues that resource users

Box 5.5 *Assessing well-being*

A goal of household surveys is to assess well-being (a livelihood outcome) through establishing total incomes. Understanding local perceptions of wealth can be critically important for researchers who seek to understand the interactions between income, wealth, resource use and well-being. Quantitative surveys and census data typically provide information on predefined notions of household wealth, making a comparison with local wealth notions important for evaluating the validity of survey results. Local poverty and wealth notions can include a high level of detail, involving more than seven progressive stages (Krishna, 2006). Perceptions of wealth differences in a community, as well as local indicators and criteria for wealth and well-being, can be identified using wealth ranking techniques (DFID, 2000). Adding a temporal dimension by conducting historical wealth rankings, possibly followed by gathering individual household trajectories, can yield information on pathways into and out of poverty realized in the community – information that will complement results from a snapshot survey. Local well-being notions may also be explored through coding qualitative interviews regarding respondents' happiness with their life and reasons for this (Camfield and Ruta, 2007).

themselves feel are important is key to understanding the context you are working in. Living in the research site for an extended period can be invaluable in building a nuanced insight and knowledge of the area. At the end of the day, the more contextual information you have, the more comfortable and confident you will be in explaining your findings from the quantitative survey and in understanding the role of natural resources and forests in the livelihoods of local people.

Key messages

- The SLF is a useful tool for thinking through what contextual information could be relevant to complement the formal questionnaire survey.
- Contextual information has three important uses: (a) *ex ante* to the survey as a basis for the questionnaire design and question formulation (see Chapter 7); (b) during implementation of the questionnaire survey through sensitizing the research team to local cultural and political issues that may affect the research (see Chapters 10 and 11) and; (c) *ex post* data collection to interpret findings and place these into perspective.
- Be constantly curious during fieldwork, and in a learning mode.

References

Agrawal, A. and Angelsen, A. (2009) 'Using community forest management to achieve REDD+ goals', in Angelsen, A. (ed) *Realising REDD+: National Strategy and Policy Positions*, CIFOR, Bogor, Indonesia

Agrawal, A. and Gibson, C. C. (1999) 'Enchantment and disenchantment: The role of community in natural resource conservation', *World Development*, vol 27, no 4, pp629–649

Asia Forest Network (2002) 'Participatory rural appraisal for community forest management: Tools and techniques', Asia Forest Network, Santa Barbara, www.communityforestryinternational.org/publications/field_methods_manual/pra_manual_tools_and_techniques.pdf, last accessed 5 February 2011

Balooni, K., Lund, J. F., Kumar, C. and Inoe, M. (2010) 'Curse or blessing? Local elites in Joint Forest Management in India's Shiwaliks', *International Journal of the Commons*, vol 4, no 2, pp707–728

Borrini-Feyerabend, G. (1997) *Beyond Fences: Seeking Social Sustainability in Conservation*, IUCN, Kasparek Verlag, Gland, Switzerland

Camfield, L. and Ruta, D. (2007) '"Translation is not enough": Using the Global Person Generated Index (GPGI) to assess individual quality of life in Bangladesh, Thailand, and Ethiopia', *Quality of Life Research*, vol 16, no 6, pp1039–1051

Chambers, R. (1992) *Rural Appraisal: Rapid, Relaxed and Participatory*, Institute of Development Studies, Brighton

Conway, T., Moser, C., Norton, A. and Farrington, J. (2002) *Rights and Livelihoods Approaches: Exploring Policy Dimensions*, Overseas Development Institute, London

Cundill, G. (2005) 'Institutional change and ecosystem dynamics in the communal areas around Mt Coke State Forest, Eastern Cape, South Africa', Master's thesis, Rhodes University, Grahamstown, South Africa

DFID (Department for International Development) (2000) 'Sustainable livelihoods guidance sheets', Department for International Development, www.livelihoods.org

Du Toit, A., Skuse, A. and Cousins, T. (2007) 'The political economy of social capital: Chronic poverty, remoteness and gender in the rural Eastern Cape', *Social Identities*, vol 13, no 4, pp521–540

Ellis, F. (2000) *Rural Livelihoods and Diversity in Developing Countries*, Oxford University Press, Oxford

Gough, I., McGregor, J. A. and Camfield, L. (2007) 'Theorizing wellbeing in international development', in Gough, I. and McGregor, J. A. (eds) *Wellbeing in Developing Countries*, Cambridge University Press, Cambridge, pp1–43

Hulme, D. and Shepard, A. (2003) 'Conceptualizing chronic poverty', *World Development*, vol 31, no 3, pp403–423

Kepe, T. (2003) 'Use, control and value of craft material - *Cyperus textilis*: Perspectives from a Mpondo Village, South Africa', *South African Geographical Journal*, vol 85, no 2, pp152–157

Krishna, A. (2006) 'Pathways out of and into poverty in 36 villages of Andhra Pradesh, India', *World Development*, vol 34, no 2, pp271–288

Leach, M., Mearns, R. and Scoones, I. (1999) 'Environmental entitlements: Dynamics and institutions in community-based natural resource management', *World Development*, vol 27, no 2, pp225–247

Lund, J. F. and Treue, T. (2008) 'Are we getting there? Evidence of decentralized forest management from the Tanzanian Miombo woodlands', *World Development*, vol 36, no 12, pp2780–2800

Makhado, Z. and Kepe, T. (2006) 'Crafting a livelihood: Local-level trade in mats and baskets in Pondoland, South Africa', *Development Southern Africa*, vol 23, no 4, pp498–509

Makurira, H., Mul, M. L., Vyagusa, N. F., Uhlenbrook, S. and Savenije, H. H. G. (2007) 'Evaluation of community-driven smallholder irrigation in dryland South Pare Mountains, Tanzania: A case study of Manoo micro dam', *Physics and Chemistry of the Earth*, vol 32, no 15–18, pp1090–1097

Mather, R. A. (2000) 'Using photomaps to support participatory processes of community forestry in the Middle Hills of Nepal', *Mountain Research and Development*, vol 20, no 2, pp154–161

Milton, K. (1985) 'Ecological foundations for subsistence strategies among the Mbuti Pygmies', *Human Ecology*, vol 13, no 1, pp71–78

Moser, C., Norton, A., Conway, T., Ferguson, C. and Vizard, P. (2001) *To Claim our Rights: Livelihood Security, Human Rights and Sustainable Development*, Overseas Development Institute, London

Murray, C. (2001) 'Livelihoods research: Some conceptual and methodological issues', background paper no 5, Chronic Poverty Research Centre, University of Manchester

Nori, M., Hirpa, A. and Ferrari, G. A. (1999) 'Complementary methods to understand land-use changes: An example from the Ethiopian Rift Valley', PLA Notes pp16–20, www.planotes.org/documents/plan_03504.PDF, last accessed 5 February 2011

North, D. C. (1990) *Institutions, Institutional Change and Economic Performance*, Cambridge University Press, Cambridge

Ribot, J. C. and Peluso, N. L. (2003) 'A theory of access', *Rural Sociology*, vol 68, no 2, pp153–181

Richards, M., Davies, J. and Cavendish, W. (1999) 'Can PRA methods be used to collect economic data? A non-timber forest product case study from Zimbabwe', PLA Notes pp34–40, www.planotes.org/documents/plan_03607.PDF, last accessed 5 February 2011

Rowley, J. (1999) 'Tips for trainers: Matrix ranking of PRA tools', PLA Notes pp47–48, www.planotes.org/documents/plan_03609.PDF, last accessed 5 February 2011

Scoones, I. (1998) 'Sustainable rural livelihoods: A framework for analysis', IDS working paper no 72, Institute of Development Studies, Brighton

Scoones, I. (2009) 'Livelihoods perspectives and rural development', *Journal of Peasant Studies*, vol 36, no 1, pp171–196

Shackleton, S. E., Campbell, B., Edmund, D. and Wollenberg, E. (2002) 'Devolution and CBRNM: Creating space for local people to participate and benefit?', *ODI Natural Resource Perspective Series*, no 76, March 2002

Sithole, B. (2002) *Where the Power Lies: Multiple Stakeholder Politics Over Natural Resources: A Participatory Methods Guide*, Center for International Forestry Research, Indonesia

Theis, J. and Grady, H. (1991) *Participatory Rapid Appraisal for Community Development*, International Institute for Environment and Development, London

Chapter 6

The Division of Labour Between Village, Household and Other Surveys

Pamela Jagger and Arild Angelsen

Not everything that can be counted counts, and not everything that counts can be counted.
William Bruce Cameron (1963, *Informal Sociology: A Casual Introduction to Sociological Thinking*, Random House, New York)

Introduction

After formulating interesting, clear and answerable research questions with associated testable hypotheses, the next task is to select the best methods for data collection. Data collection aims to obtain the most accurate and precise measures of variables of interest (see Chapter 11). The challenge is to maximize data *validity* and *reliability* (see Box 3.2 for definitions) given the constraints of research budgets, researcher and respondent time, and the willingness and capacity of respondents to answer the types of questions included.

Researchers have a diverse set of methods to choose from. In this chapter we cover approaches to collecting village and household-level data, using rigorous qualitative and quantitative methods. The two major points of this chapter are: (a) think carefully through the nature and use of data, and (b) choose the scale and format for data collection based on that. As part of this, consideration should be given to when a quantitative indicator is needed to explore the research question. Some questions and data collection efforts lend themselves to more qualitative approaches.

Before reviewing the main survey approaches, a reminder of the main uses of survey data is in order. There are three types of information a field researcher should collect:

1. Data for the quantitative (statistical) analysis: The title of this book – *Measuring Livelihoods and Environmental Dependence* – points to a focus on the quantitative analysis. Specifying the exact data that are needed to answer the research question and test the hypotheses is a critical element of fieldwork preparation. The data needs have – hopefully – been identified as part of the research proposal and matrix (Chapter 3), but this is a continuous task until data collection starts.

2. Background (contextual) information: Background information will not have the same requirements for representativeness, exact definitions and specification as the quantitative data. Still, it is essential to provide background for in-depth study, partly to enable the interpretation of statistical analyses (see Chapter 5).

3. Information to situate the study area in a broader context: Data on larger scale structural variables help to situate the analysis in the broader context of the sub-national or national landscape, and inform about the generalizability of the study. A research finding that claims to be representative of 15 million people is much more interesting than one representative of only the 1500 in the study villages. Ideally, one should be able to say how representative the study areas and sample population are of the sub-national (for large and diverse countries such as Brazil or Indonesia) or national context. Typical variables useful for addressing the issues of generalizability of findings are: agroecological zone, market access, income levels, major economic activities, population density and dominant ethnic or linguistic group. For example, the research may take place in an area that is biophysically similar to 20 per cent of the land area in the country or socioculturally similar to 30 per cent of the population in the sub-national region where the study area is located. This provides the consumers of the research with important information regarding how applicable the findings are to the wider context.

Based on these different uses of data and other considerations, we shall outline four main categories of data to be collected during field research. We then elaborate on data collection at the village (community) level, followed by a brief treatment of household level surveys (covered in later chapters). Then we suggest other relevant surveys of, for example, local institutions, depending on the focus of the research. Before we conclude by stressing the need for a nested approach to data collection.

Which survey approach to take?

The household survey is the staple of most fieldwork focused on how local people utilize, manage and are affected by policies and programmes related to natural resources. There is a strong tendency to collect as much information about household demographics, socio-economic characteristics and economic decision-making as possible (in other words, income, consumption, expenditure, time use, and so on). Our experience suggests that the household survey easily becomes overloaded. Almost invariably, the pretesting experience reveals that the questionnaire is too long! One reason is that the data needs are not well defined, thus 'to be on the safe side' too many questions are included. By focusing the research questions and carefully thinking through hypotheses and possible statistical model specifications, questionnaires can be limited to include only essential data.

Another reason why household questionnaires are frequently too long is that they include information that can be more accurately and efficiently collected through other survey approaches. To determine the best methods for collecting data, the researcher should ask two key questions about every variable (or question) considered for inclusion in a survey instrument:

1. Is this variable likely to *vary* within the village/community? If *yes*, the information should be collected at the household level, if *no*, it can be collected at the village level (or higher scales).
2. Can one get *reliable quantitative figures* for this variable, and does one *need* to get representative quantitative figures for this variable to answer the research question or test hypotheses? If the answer to both parts of this question is *yes*, put the survey question in the *household* survey. If *no*, go for key informant or focus group/village discussions.

The answers to these questions enable categorization of the information needed into one of four cases (Table 6.1).

Going through this process to identify at what scale data should be collected (Q1), and whether representative quantitative data are needed (Q2) is essential to collecting the most accurate and precise data you can, and in an efficient manner.

Variation in data is a central concept in research and data analysis. Without variation in the variable of interest, there may not be an interesting story to tell and certainly very limited scope for statistical analysis. The level at which the

Table 6.1 *Matrix for deciding scale and methods for data collection*

		Q1: Does the variable vary within village?	
		Yes	**No**
Q2: Are representative quantitative figures feasible and needed?	**Yes**	Structured household survey	Structured village survey
	No	Key informants, focus groups (subset of villagers)	Village meeting

variation in the variable of interest occurs both limits the range of possible analyses and also has implications for data collection methods. In general, collect the information at the level at which the variation occurs! If all the households in a village use the same forest area, the information about that forest should be collected at the village level. There are in-between cases; market access, for example, is influenced by both the location of the village and the location of the household within the village, and may therefore be collected at both levels.

The second question of qualitative versus quantitative information is only partly related to the nature of the variable. Most variables can be measured, with varying degrees of effort and accuracy. It is easier to get a quantitative answer to the question 'how old are you?' than 'how happy are you?', but much research has gone into measuring happiness (for example the World Database on Happiness). Likewise, it is easier to measure physical capital than social capital of households, but many indicators have been developed for the latter (Pretty and Ward, 2001; Katz, 2000; Gibson et al, 2005).

Thus, equally important as the nature of the variable is the need of the research project and how the information will be used. Is it to be used as background information or in statistical analysis? Using the table will typically result in a nested approach. Different variables of the same topic area are included in different surveys. For example, if land rights and tenure are important in the research project, information regarding the history of tenure in the community can be collected through key informant interviews, major problems and land conflicts can be on the agenda in a village meeting, while the household questionnaire may contain questions about the household's land-ownership and involvement in land conflicts, for example, to test a hypothesis regarding the poor being more vulnerable to land conflicts.

In our experience, the main benefit of this approach is that it results in a well-specified and parsimonious household-level questionnaire. It almost invariably results in a shorter questionnaire, leading to better quality data, reducing fieldwork costs, and minimizing the burden on the respondents. Conversely, the danger of this approach is that too much information will be integrated into village-level survey instruments. It is essential to consider how the data collected will be used and analysed; this rule applies to questions asked at both the village and household scales, and for both qualitative and quantitative data.

Collecting village data

This section and the one that follows will provide a brief overview of the main survey instruments outlined in Table 6.1. To avoid overlaps with Chapter 5 regarding contextual information and the chapters that follow on household surveys, the coverage is somewhat uneven, with a stronger emphasis on village surveys as these are not covered elsewhere.

What data to collect

Careful thought should go into designing village-level surveys. As we indicated above, village surveys frequently become too long as they serve as a catch-all for questions that did not make it into the household survey. Long village surveys are difficult to administer in the field as they require participants to devote too much of their time. Thus, the critical question is again: how are the data to be used – as general background and contextual information or for statistical analysis?

The structured village questionnaire

Village questionnaires can have both qualitative and quantitative components. Data to be used in the core statistical analysis need to be collected as quantitative information; to collect these data across a diversity of villages a structured village questionnaire should be developed.[1] Suggested data include:

- Geography (global positioning system (GPS) location of village, average rainfall, trends in rainfall, altitude, slope).
- Demography (population, in-migration and out-migration, ethnicity).
- Infrastructure (water and electricity sources; presence of education and health facilities).
- Land uses.
- Forest resources (distance to nearest forest; biophysical condition; most important products harvested).
- Forest institutions (property rights; forest user groups).
- Shocks and crises that put households at risk (for example, drought, fire, war).
- Wages and prices.

These data can be used in several ways. They can be used to group villages into clusters, for example, four clusters based on location (remote–central) and local forest management institutions (weak–strong). If the number of villages is large, data can be used in a regression analysis of households' forest use, including variables such as distance to forest, number of forest user groups, and so on, rather than more general village dummy variables (that are often hard to interpret).

Background and contextual information

There are various types of information that may not be included in the statistical analysis but still contribute significantly to the analysis. Such data can provide important background and contextual information, help to reformulate and make more explicit hypotheses, help to construct or impute data, assist in interpreting statistical results, rule out alternative explanations for your findings and situate the study in the sub-national or national context.

The following are examples of data to collect:

- History of village.
- Main livelihood activities.
- Seasonal calendars for agriculture and forest products.
- Seasonal and/or historical price data for major agriculture and forest commodities.
- Dates and effects of current and past major political, economic, biophysical, weather, in-migration and out-migration events.
- The quality of public services, including roads, schools, health centres, water sources, and so on.
- Information regarding important social and cultural aspects of society (for example, marital norms or gender roles).
- Narratives of major drivers of land use and environmental change.

We recommend that researchers take a systematic approach to collecting background information if there is more than one village included in the study. Having a complete set of information for each village in the study area allows for more rigorous analysis and explanation of observed phenomena. Researchers should collect and approach the analysis of qualitative data with the same rigour as they would approach quantitative data. This may involve coding qualitative data, undertaking content analysis and creating typologies that situate villages according to important information.

How to collect village-level data

There are a variety of ways to collect village-level data, and filling out a village questionnaire typically involves several methods of data collection. Whenever possible, the researchers should rely on their own observation/measurement and reliable secondary data sources, rather than burdening focus groups, key informants or village meetings with unnecessary questions. Time spent with village members should be spent capturing data that is not available from other sources. There are six possible sources of information for village-level variables: data collected using own observation or measurement; secondary sources; village officials; key informants; village meetings or focus groups; and village census.

Own observation and measurement

Some data can be captured by own observation or by taking measurements using a vehicle odometer or GPS. Quite a lot can be observed about a village simply by spending time there. Data on the presence of most physical infrastructure – including schools, health centres, boreholes, and so on – can be collected by observation, negating the need to include questions about infrastructure in a village-level meeting. Measurements such as distance to the nearest all season road, nearest forest, area of village, altitude, and so on, can be measured and recorded on the village questionnaire by the researcher.

Secondary data

Information about the village or region might be provided through national or regional statistical yearbooks, census reports, statistical bureaus or large-scale surveys undertaken in collaboration with bilateral or multilateral institutions (for example, the Living Standards Measurements Survey at the World Bank, see Box 7.4). Reliable records should be used whenever possible to capture basic demographic and public service data. Care should be taken to document secondary data and reference it accordingly. Time lags between the collection of data and the processing and reporting of data in developing countries means that data might be several years out of date. Researchers should be cautious about the reliability and relevance of data depending on the source, when the data were collected and whether data are disaggregated to a level where they are indicative of conditions in study villages. Many villages also have good records of population and in- and out-migration from the village, access to public services, land categories, and so on.

Several types of data that are not available at the village level might be available at higher administrative levels or at the landscape level. For example, rainfall may be an important variable for explaining variation between study villages. Rainfall data is generally collected and recorded by national or sub-national government authorities, or possibly NGOs, and often documented and easily available in some centralized location, for example at the Ministry of Environment. If not, visit rainfall data collection points (for example, weather stations, airports and airstrips, local colleges or district headquarters) and compile data. Other types of data may be available at higher levels of administration than the village including local government spending on forestry and agricultural extension, vaccination rates and educational attainment.

Village officials

Village officials can be an excellent source of factual information. When reliable written records are not available, village officials may have some of the factual information needed. For example, village leaders who hold the right to allocate land should be able to state very accurately the number of in-migrant households over a given time period. Similarly, village officials may also have time series data for the population in the village, which can be very useful to complement an oral history.

It may be more efficient to ask village leaders these types of questions rather than to put them to a larger group where time typically is needed for aggregating the collective knowledge. However, we caution against relying only on village leaders for answering questions involving subjective assessment. Their responses may be a biased or 'polished' view of the state of affairs. For example, asking a village leader about land conflicts in the village may yield an incomplete response. The leader may not want to discuss problem issues in the village, or may himself be involved in a land conflict.

Key informants

Key informants are residents of the village that have a high level of awareness regarding social, economic, demographic and cultural trends. They are frequently politically active and engaged in governance either formally or informally. They may have lived in the village longer and held key positions. They are typically more curious about village affairs, and so on. Key informant interviews are generally more informal than focus group meetings, but one should still have an interview guide, that is, a set of questions to be discussed. Key informants may be among the respondents in a household survey. It is important to always keep a list of interesting questions, not suitable for the formal questionnaire, and ask households that seem particularly well-informed.

When interviewing key informants, it is important to beware of biases: a seemingly very well-informed key informant might have a biased view for some reason. For example, he or she represents one group in the village or they may want to hide certain information to portray their village positively. As a general rule, triangulate information from key informants, ask many people the same questions and eventually the answers will converge towards a more complete picture. Key informants can also be important sources of sensitive information, but this requires a more relaxed atmosphere (no pen and paper or microphone, but write down the information as quickly as possible after the conversation has ended).

Valuable data can be collected by sitting with community members and having informal discussions. These types of interactions are generally unscripted, providing an opportunity for community members to talk about things that might be outside of the scope of the core research instruments. Consider asking questions after a game of chess or during the village market day. Spend time with respondents on informal terms and discuss controversial or sensitive issues (for example, if illegal timber is being harvested in the forest).

While informal discussions take place without questionnaires in hand, take detailed notes during the discussion. At a minimum, collect details on the name of the person interviewed, the date and where/how they can be contacted for follow-up questions. There is a good chance that these types of interactions will lead to notebooks full of interesting information that will never make it into the final research outputs. However, informal interviews can reveal important details that help with focusing on key variables or motivating new lines of inquiry.

Village meetings and focus groups

It is strongly recommended to have at least one focus group or village meeting to gather important qualitative information. Such meetings are essential for collecting data that involves some degree of subjectivity. A village or focus group meeting can be organized in different ways. Depending on the size, one alternative may be to invite all village members. Another option is to call a smaller group, say eight to ten people. In many countries, it is not just expected but mandatory to go through the village leadership when organizing such meetings. Not doing so can be seen as both impolite and possibly also a direct violation of the rules and regulations and can seriously obstruct the research. But one should also be aware that a village leader may select an unrepresentative group, thus one should ask the village head to invite a diverse group (men and women, young and old, rich and poor, immigrants and long time residents, and so on). During the village/focus group meeting have a list of questions ready and ask respondents in a systematic way. Follow up on interesting leads, but do not get sidetracked. If research is conducted across several villages, a semi-structured or structured village questionnaire should be developed to collect information in a systematic and comprehensive manner. In general, village meetings or focus groups should not last for more than two hours. Participants may spend a lot of time waiting for the group to assemble prior to the formal start of the discussion. Always respect the time constraints of respondents.

If contradicting views and information occur during the meetings, try to reach a consensus answer. More generally, it is critical to double-check information given by individuals. Thus, ask the same question to many individuals. This is particularly important for information that potentially could be sensitive, controversial or particularly important for answering the key research questions.

The lead researchers should be present during village meetings and be responsible for filling in the village questionnaires – this task cannot be delegated. This is unlike the household questionnaire where, after an initial training period, enumerators can do much of the data collection. The reason is that the information in the village questionnaires requires a more critical assessment and judgement. Fill in as much as possible while conducting the survey and have an enumerator also taking notes and collecting important information that comes out of side discussions related to the questions. Having two people recording information will help with capturing as much of the rich discussion as possible. Compensate village/focus group members for their time by providing a small snack and/or drink during the meeting.

Village census

A village census can be used to collect accurate demographic information for the village questionnaire and can be a useful instrument for two main reasons. It can provide important data on basic demographic variables (number of household members, age/sex/education of household head, caste/ethnicity, in-migration) key livelihoods activities (for example, main occupation) and other areas of interest. But, as the census is to cover all the households in the village, it has to be very brief, often limited to ten key questions that can easily be answered.

The other main reason for undertaking a village census is to serve as your sampling frame for selection of households to be interviewed with the household questionnaire. The complete list of households can be used to randomly draw the chosen number of households. But, a census would be even more useful in stratification if that procedure is chosen (Chapter 4). For example, if the research focuses on a particular forest product, which is collected by only a minority of the households, the census can identify those households. The sampling procedure might then be to select equally sized samples of collectors and non-collectors, and use the census result in a weighted aggregation to generate representative village data.

Undertaking village census can be time-consuming, particularly in villages with large populations (more than 500 households) and covering large geographical areas. Thus, a cost–benefit analysis is needed. If the purpose is just to get a list for the random sampling for the household survey, other methods are likely to be more efficient (see Chapter 4).

Household surveys: Structured formal quantitative questionnaires

Household surveys in the fields of agricultural, resource, environmental and development economics are generally focused on the collection of quantitative data that can be used in statistical models specified to explain household level behaviour. Questionnaire design should be focused on capturing all of the data required for the behaviour model specified. The possibility of obtaining accurate quantitative responses, and variation across households in responses to specific questions, are the critical elements in designing household questionnaires. Household surveys should only include questions that elicit data expected to vary from household to household. 'Has your household been negatively affected by a recent drought?' is not a good question for a household survey. Droughts generally manifest as a covariate shock, meaning that all households in the drought-affected region are similarly impacted. However, asking households about coping mechanisms related to the recent drought is a good question for a household survey.

A broad discussion of household questionnaire design is given in the next chapter. One issue concerns the frequency of surveys, linked to the accuracy and precision of household responses to questions. Frequently, researchers seek to explain economic behaviour within the household for a period of a year. Annual data are important as they reflect seasonal variation, and are comparable with other standard statistics produced at the sub-national and national level. Most household-level socio-economic surveys are administered one time only, meaning that the researcher has to come up with creative ways to elicit accurate information from household respondents. This can be a serious challenge as it is very difficult for most of us to recount our full income, consumption, expenditures or time-use portfolios for the past week, much less the past year. We have a limited cognitive ability to recall, with any degree of accuracy, over a long period of time. Particularly challenging to recall and aggregate are regular transactions or events; irregular economic activity, such as expenditures for a wedding, are easier to recall with accuracy. Researchers have developed a variety of ways to deal with the recall issue including: administering questionnaires multiple times to capture seasonal variation or to parse the year into smaller units of recall and aggregation (for example, the Poverty Environment Network (PEN); Campbell et al, 2002; Nielsen and Reenberg, 2010); providing variable recall periods depending on the type of product or activity to be quantified (Cavendish, 2002), for example, more regularized activities should have shorter recall periods; or using participatory methods rather than more formal accounting methods to motivate households to think about the relative rank and weight of various livelihood strategies (Box 6.1).

Other livelihood related surveys

Beyond village and household questionnaires, there are other types of data to consider collecting to supplement, complement or triangulate the rich data collected using the core village and household research instruments. For example, other types of data include: additional data that provide finer detail or

Box 6.1 *Participatory techniques versus detailed accounting approaches: Do the methods matter?*

Pamela Jagger, Marty Luckert, Abwoli Banana and Joseph Bahati

Virtually every researcher engaged in social science fieldwork has faced decisions regarding using aggregated or disaggregated approaches to collect information. These decisions are made largely based on individual experiences and disciplinary training. We conducted an experiment in Uganda to test whether different methods of data collection yield significantly different results. We collected information on rural income portfolios for two sub-samples of the same population of households in western Uganda using different survey instruments: a highly disaggregated income survey, and a participatory rural appraisal survey instrument that collected household-level information using a more aggregated approach. For example, in the disaggregated household survey, respondents were asked to itemize forest products harvested and to indicate the quantity and value of the products, as well as any financial costs incurred in their production. By combining these data with estimates of net income for other sectors of the livelihood portfolio, we were able to estimate the share of the total portfolio from forest products as well as for other important sources of income. Conversely, the aggregated approach involved asking households to rank and weigh ten categories of income by coming to consensus about appropriate rankings and weighting. Using this participatory method, we were also able to estimate income portfolio shares. We then compared the results of these two approaches to see whether and why they are different (Figure 6.1).

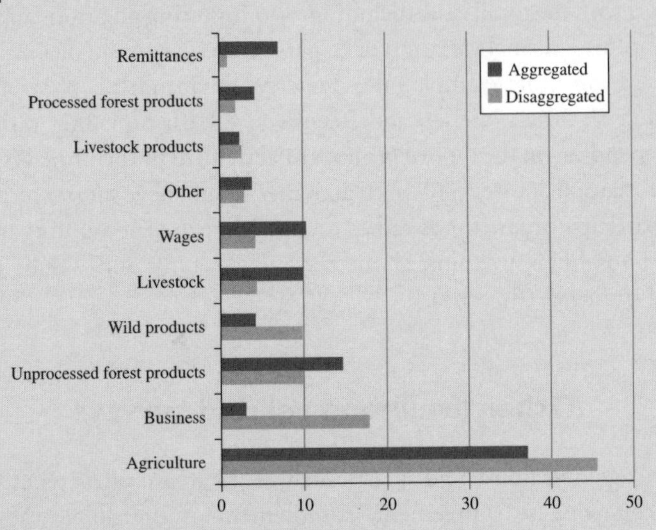

Figure 6.1 *Income portfolio shares*

The findings demonstrate that different data collection approaches yield significantly different results. The disaggregated data indicate that agriculture, business and unprocessed forest products are the three most important sources of household income. Using the aggregated method, wage income replaces business income in the top three. When we look at the overall distribution of the shares for each method, we observe a smoother distribution of income among the various categories for the aggregated data collection. We have no way of knowing which of these methods is most accurate. The observed differences in the data collection methods motivate us to consider what factors household respondents consider when responding to survey questions. The aggregated approach requires respondents to think holistically about the relative importance of the various income sources, including factoring in the activities of all household members over the calendar year. The disaggregated approach requires household respondents to reconstruct all income-related activity for all household members during a shorter time period.

fill in missing values in village and household surveys (for example, price and wage surveys); informal discussions that add contextual details and new background information; focused surveys on specific topics that require different sampling strategies and questionnaires (for example, value chain surveys); and data that need to be collected at a wider scale than the village or are more appropriately linked to biophysical boundaries rather than the political boundaries of villages (for example, local forest or water user groups).

Price and wage surveys

Additional surveys can be undertaken to systematically collect data that may be important for imputing values that are missing from the data set. For example, households that collect but do not sell fuel wood may find it difficult to indicate a value for a head load of fuel wood. If fuel wood is sold by other households in the village or by traders that come to the village, it should be possible to obtain a village-level price for fuel wood. This value can be used to calculate the economic value of fuel wood to households that were unable to provide price information (see also Chapter 8).

Daily wage rates for adult male, adult female and child labour, and village-level price data for agriculture, livestock, forestry and other environmental products can be collected using village-level focus groups. Focus groups should be comprised of representative groups from the village, including men and women of varying ages, socio-economic status and ethnic groups. Village trading centres or markets are good places to find a group of people to interview. It is a good idea to have a very short questionnaire that allows systematic entry of these data. When collecting price data, care should be taken to make sure that consistent units and conversion factors for products that are sold in a variety of different units are used.

Also take note of seasonal change in wages and prices. Researchers should collect wage and price data for at least two seasons, the busy and slack agricultural periods of the agricultural calendar. If the field research lasts for a prolonged period of time, interesting seasonal price fluctuations can also be revealed by regular (for example, weekly) collection of market prices for key commodities.

Value chain surveys

Researchers interested in forest and environmental resources may undertake value or commodity chain studies for forest products that are important to rural livelihoods (for a review of methods see Kaplinsky and Morris, 2001; Ribot, 2005). Depending on the focus of the study, it may be decided to collect detailed data on value chains for specific products. Generally those that are highest in value, have reasonably robust markets and contribute the most to forest-based livelihoods are of interest. Various aspects of value chains can be studied using value chain surveys, including: profits and marketing margins across a diversity of value chain participants; producer groups and associations functioning in the area where the study is taking place; rules, regulations, taxes and fees pertaining to the production, trade, transport and retailing of products; the availability and volume of credit to value chain participants, and so on. As with household and village-level data, these types of data should be collected using rigorous methods including identifying a representative sample or surveying the relevant population, and by developing research instruments with well thought-out questions accompanied by appropriate recording and coding structures. Examples of studies focused on sub-national value chains include: Ribot (1998, charcoal), Gellert (2003, timber), Shively et al (2010, charcoal).

Local institutions and groups

The study of local institutions and their relationship to forest management and livelihood outcomes has shed light on the importance of studying collective action for sustainable forest management and resource use (Varughese and Ostrom, 2001; Adhikari, 2005; Agrawal and Chhatre, 2005; Jumbe and Angelsen, 2006, 2007). Participation in local organizations such as forest user groups, microlending groups, and so on, can be important determinants of household-level dependence on forest and environmental resources. Household surveys should include data on household participation in such groups. In study areas where groups play a significant role in shaping resource use, a separate questionnaire focused on local institutions and groups can be implemented. The International Forestry Resources and Institutions (IFRI) research programme has been collecting data on forest governance and institutions for more than 15 years (Ostrom and Wertime, 2000).[2] Their resources provide an excellent starting point for developing questionnaires focused on local institutions, collective action and forests. Such additional surveys may require a different sampling procedure than the one used for village and household-level data collection. For example, an independent survey on forest user groups might involve all members that use products from a specific forest landscape.

Sub-populations

Understanding how various sub-populations utilize environmental resources is frequently of interest. This requires data that allows for disaggregating a sample of households by sub-population. Households can be split into sub-populations representing different groups, including: female-headed households; migrant households; ethnic minority households; relatively poor households, and so on. Calculating basic descriptive statistics and/or running regressions using split samples provides interesting insights into how various groups are differentially affected by changes in natural resource management policies, property rights, land tenure etc. See Jagger (2008) for an example of a split sample multivariate analysis examining the impact of Uganda's forest sector reform on the relatively wealthy and relatively poor households.

Data can also be collected at the individual level. Intra-household surveys are the best way to learn how women, men, youth, elderly and sick members of households utilize resources or are differentially affected by policies or projects (Haddad et al, 1997; Maggs and Hoddinott, 1997; Sapkota and Odén, 2008; de Sherbinin et al, 2008). Intra-household surveys involve interviewing several household members to ascertain differences that exist across gender and age cohorts. As with our advice on variation above, intra-household data should only be collected for variables likely to vary among household members.

Conclusions – a nested approach

The research strategy is central to the development of household and village-level questionnaires, and the field researcher needs to keep in mind the overall research questions and the testing of alternative hypotheses. The research process, including choice of survey instruments, is influenced by that. The key message of this chapter is to adopt a nested approach, where the information needed is collected at the appropriate scale (in other words, the level where the variation in the variable occurs) and, depending on the later use of that information, is collected using qualitative or quantitative data collection methods. This can be achieved by asking two very simple questions, as outlined in Table 6.1. It has large benefits in terms of more concise and cost-efficient research instruments. Yet, there is always a risk of overloading the questionnaires; in particular the structured village questionnaire might become a dumping site for questions that did not make it into the household questionnaire.

We realize that our categorical view of qualitative versus quantitative data is rather black and white. Empirical field research on social processes almost always has more nuance to it; researchers need to be flexible in their approach. In addition, the need for triangulation using different types of survey questions to elicit the same information suggests the use of both qualitative and quantitative data to ensure robust analysis and research findings. Developing additional questionnaires on prices, wages, value chains, local institutions and sub-populations can facilitate triangulation.

Key messages

- Variation is essential for a robust analysis: think about at what *scale* you expect to see variation in your data (for example, household, village, sub-national or landscape levels).
- Think carefully about whether you need qualitative or quantitative data to answer your research question.
- Use a variety of methods at multiple scales to triangulate data for the most important variables.

Notes

1 The Poverty Environment Network (PEN) village questionnaire illustrates the type of information that can be collected (see www.cifor.cgiar.org/pen/_ref/tools/prototype.htm for a prototype questionnaire).

2 See http://sitemaker.umich.edu/ifri/resources for an overview and links to research methodology, instructions and research instruments from the IFRI research programme.

References

Adhikari, B. (2005) 'Poverty, property rights and collective action: Understanding the distributive aspects of common property resource management', *Environment and Development Economics*, vol 10, no 1, pp7–31

Agrawal, A. and Chhatre, A. (2005) 'Explaining success on the commons: Community forest governance in the Indian Himalaya', *World Development*, vol 34, no 1, pp149–166

Campbell, B. M., Jeffrey, S., Kozanayi, W., Luckert, M., Mutamba, M. and Zhindi, C. (2002) *Household Livelihoods in Semi-Arid Regions: Options and Constraints*, Center for International Forestry Research, Bogor, Indonesia

Cavendish, W. (2002) 'Quantitative methods for estimating the economic value of resource use to rural households', in Campbell, B. M. and Luckert, M. (eds) *Uncovering the Hidden Harvest: Valuation Methods for Woodland and Forest Resources*, Earthscan, London, pp17–65

de Sherbinin, A., VanWey, L. K., McSweeney, K., Aggarwal, R., Barbieri, A., Henry, S., Hunter, L. M., Twine, W. and Walker, R. (2008) 'Rural household micro-demographics, livelihoods and the environment', *Global Environmental Change*, vol 18, no 1, pp38–53

Gellert, P. K. (2003) 'Renegotiating a timber commodity chain: Lessons from Indonesia on the political construction of global commodity chains', *Sociological Forum*, vol 18, no 1, pp53–84

Gibson, C. C., Williams, J. T. and Ostrom, E. (2005) 'Local enforcement and better forests', *World Development*, vol 33, no 2, pp273–284

Haddad, L., Hoddinott, J. and Alderman, H. (eds) (1997) *Intrahousehold Resource Allocation in Developing Countries. Models, Methods, and Policy*, The Johns Hopkins University Press, Baltimore, MD and London

Jagger, P. (2008) 'Forest incomes after Uganda's forest sector reform: Are the poor gaining?', CAPRi (CGIAR System Wide Program on Collective Action and Property Rights) Working Paper Series No 92, International Food Policy Research Institute, Washington, DC

Jumbe, C. B. L. and Angelsen, A. (2006) 'Do the poor benefit from devolution policies? Evidence from Malawi's forest co-management program', *Land Economics*, vol 82, no 4, pp562–581

Jumbe, C. B. L. and Angelsen, A. (2007) 'Forest dependence and participation in CPR management: Empirical evidence from forest co-management in Malawi', *Ecological Economics*, vol 62, no 3–4, pp661–672

Kaplinsky, R. and Morris, M. (2001) *A Handbook for Value Chain Analysis*, International Development Research Centre, Ottawa, Canada

Katz, E. G. (2000) 'Social capital and natural capital: A comparative analysis of land tenure and natural resource management in Guatemala', *Land Economics*, vol 76, no 1, pp114–132

Maggs, P. and Hoddinott, J. (1997) 'The impact of changes in common property resource management on intrahousehold allocation', FCND Discussion Paper 34, International Food Policy Research Institute, Washington, DC

Nielsen, J.Ø. and Reenberg, A. (2010) 'Cultural barriers to climate change adaptation: A case study from Northern Burkina Faso', *Global Environmental Change*, vol 20, no 1, pp142–152

Ostrom, E. and Wertime, M. B. (2000) 'International forestry resources and institutions research strategy', in Gibson, C. C., McKean, M. A. and Ostrom, E. (eds) *People and Forests: Communities, Institutions, and Governance*, MIT Press, Cambridge, MA, pp243–268

Pretty, J. and Ward, H. (2001) 'Social capital and the environment', *World Development*, vol 29, no 2, pp209–227

Ribot, J. C. (1998) 'Theorizing access: Forest profits along Senegal's charcoal commodity chain', *Development and Change*, vol 29, no 2, pp307–341

Ribot, J. C. (2005) 'Policy and distributional equity in natural resource commodity markets: Commodity-chain analysis as a policy tool: A research concept paper', World Resources Institute, Washington, DC

Sapkota, I. P. and Odén, P. C. (2008) 'Household characteristics and dependency on community forests in Terai of Nepal', *International Journal of Social Forestry*, vol 1, no 2, pp123–144

Shively, G., Jagger, P., Sserunkuuma, D., Arinatwe, A. and Chibwana, C. (2010) 'Profits and margins along Uganda's charcoal value chain', *International Forestry Review*, vol 12, no 3, pp270–283

Varughese, G. and Ostrom, E. (2001) 'The contested role of heterogeneity in collective action: Some evidence from community forestry in Nepal', *World Development*, vol 29, no 5, pp747–765

Chapter 7

Designing the Household Questionnaire

Arild Angelsen and Jens Friis Lund

Differences in question design often bring results which show far greater variation than those normally found by different sampling techniques.
George Gallup (1947, cited in Foddy, 1993)

'What was your forest income last year?'

Imagine a researcher asking a farmer in Indonesia the following question: 'What was your forest income last year?' This is a poor question to ask because the concepts are not clear. For example, (a) does 'your' hint to the person or the household? (b) Are 'forest' products those from a small woodlot on the farm or a deer roaming in the forest during the daytime and feeding on farmland during the night? (c) Does 'income' imply cash income only (as many will associate 'income') or cash and subsistence income combined (the economic definition of 'income')? And (d) does 'last year' refer to the past 12 months or the last calendar year (and which calendar, by the way)?

Even if the farmer understood the question to be: 'What was the cash and subsistence income for your household from forest X over the past 12 months?', he would probably not have a clue about what the income was. The household's forest income might consist of 20 different products, be collected by five different household members during four different seasons, and mainly be used for family consumption (with the price unknown). In other words, the question begs of the farmer a large algebraic operation with several unknown variables! If the farmer does come up with an answer – if nothing else as a display of politeness – it is unlikely to reflect the household's *actual* forest income last year.

This chapter gives guidance on questionnaire design and question formulations that will increase the quality of the data collected. In the case of our Indonesian farmer, it means asking carefully worded questions in an appropriate sequence so that answers provide a measure of our farmer's forest income – and other relevant variables – that is close to reality. Our focus is on the household questionnaire. Although this has become the staple of quantitative fieldwork, designing a questionnaire is not a quick and easy way to collect empirical data. As just seen, even for well-defined quantitative concepts such as 'income', designing questions that facilitate the collection of valid data is not a simple task. Any short cuts taken in the questionnaire design phase will almost invariably have to be paid back with high interest at later stages of the research.

While this chapter focuses on the questionnaire design, it presumes that the homework has been done. The research questions, hypotheses and data needs should have been sorted out (Chapter 3), the sampling strategy elaborated (Chapter 4) and an overview of the surveys suitable to gather the different data should be available (Chapter 6). The gross list of data to be produced by a survey then needs to be reformulated into questions that can be asked in a questionnaire. The 'translation' process – from data needs to question formulation and questionnaire design – entails several risks: (a) dropping – some data needs are not translated into questions, (b) modification – questions do not reflect data needs accurately (this has serious implications for construct validity, see Box 3.2 for a definition of construct validity), (c) duplication – unnecessary costs are introduced as the same data needs are inquired about in more questions, and (d) expanded scope – questions are introduced that elicit data not needed for the analysis. Every unnecessary question in the final questionnaire will be extremely costly in terms of the time needed for design, testing, collection, entry, checking and, subsequently, analysis of data that is not needed to answer the research questions – as well as the negative effect it may have on the quality of other data collected by dragging out the interview. Thus, each question that goes into the questionnaire must pass the test: how will this information be used? Obviously, this is an iterative process, where it will be necessary to go back and forth between the research questions and hypotheses and the questionnaire to ensure alignment (see Box 7.1 for types of variables).

The basic building blocks: HAI+

Many household questionnaires follow the HAI + format, implying that they include the following four groups of questions:

- **H**ousehold composition and characteristics.
- **A**ssets owned by the household.
- **I**ncome of the household.
- + special sections reflecting the particular focus of the research project.

Box 7.1 *Types of variables*

Depending on how questions are formulated, the household questionnaire can generate a number of different types of variables (de Long, 1997):

- **Continuous variables** are any value within a defined range of values, such as prices, volumes and weights.
- **Censored variables** occur when all values in a certain range are reported as (or transformed to) a single value – typically the dependent variable is zero – for a large proportion of the observations. For example, households that do not harvest any forest products will necessarily have zero forest income.
- **Binary variables** have two categories and are often used to indicate that an event has occurred or some characteristic is present. For example: does the household belong to a forest user group? 0 = No; 1 = Yes.
- **Ordinal variables** have categories that can be ranked. For example: how has forest cover in the village changed during the past five years? 1 = Major decline; 2 = Minor decline; 3 = No change; 4 = Minor increase; 5 = Major increase.
- **Nominal variables** occur when there are multiple possible outcomes or responses that cannot be ordered. For example: what is the primary livelihood activity of your household? 1 = Agriculture; 2 = Livestock; 3 = Forestry; 4 = Wage labour; 5 = Business; 6 = Other.
- **Count variables** indicate the number of times an event has occurred. For example, how many days did you harvest bush meat during the past year? Possible responses range from 0 to 365.

As you formulate your questionnaire you need to think carefully about what type of quantitative data you want to collect. How you ask questions has implications for the type of analysis that you can undertake after you have collected and cleaned your data (see Chapter 13). Some types of information are clearly best collected as continuous variables – for example, the distance from the house to the edge of the forest.

For variables related to education, you could think about collecting data in a variety of ways. For example, data on education can be collected in the following ways:

- How many years of education did the household head complete? (Censored, 0 through approximately 20)
- What is the highest level of education completed by the household head? (Ordinal, 0 = None; 1 = Some primary; 2 = Completed primary; 3 = Some secondary; 4 = Completed secondary; 5 = Beyond secondary)
- Did the household head complete primary school? (Binary, 0 = No; 1 = Yes)

If you are uncertain as to how you would like to specify your variables, you might consider collecting data at the highest level of practical detail. In the education example, you would collect data on education as a continuous variable as this would allow you the flexibility to create ordinal or binary variables later.

Household composition and characteristics

Household composition and characteristics include basic information about the household members, such as age, sex, level of education, kin (relation to household head) and main occupation. A table with these characteristics as columns and individual household members as rows is typically the opening part of the questionnaire. Relatively detailed information about household composition is necessary for investigation of intra-household patterns of income earning and/or consumption, but is also relevant if household income is to be compared across households. In the latter case, data on age and sex can be used to compute an adult-equivalent household size or similar measures of household standardization (see Cavendish (2002) for an example of how to compute this).

Further, the H-section might include basic information about the household: when was it established, when did they move to the village, ethnicity/caste, religion and location relative to the village centre, rivers, forests and other points in space of interest to the research project.

Assets

Assets are productive resources that can be accumulated, sold or invested to achieve desired ends (livelihood outcomes). The assets that the household own or have access to form the basis for the generation of household income, thus the 'asset-income' pair is analogous to the 'stock-flow' of natural sciences. Assets influence households' choice of livelihood strategies and are used as indicators of wealth that are less dependent on annual fluctuations than income. They are therefore almost invariably included in livelihoods and income surveys.

The asset recording should include estimates of all of the five types of assets or capital in the Sustainable Livelihoods Framework (SLF): natural, human, physical, social and financial capital (see Chapter 5). The five types of capital assets differ along two important dimensions, ownership and tangibility, which have implications for the way they are measured.

Property rights to natural and physical assets are often ambiguous in rural areas of developing countries (see Barzel, 1989; Bromley, 1991). In general, ownership or property rights are a bundle of rights, the two major ones being the *use* (including current income) rights and the *transfer* rights (including sell or rent). An operational definition of property rights that include the *transfer right* is appropriate in relation to financial and many physical assets, for example, cash, jewellery, house and livestock. For some physical assets (such as collectively owned agricultural machinery or irrigation systems) and natural capital (for example, grazing land) this can be too limiting. Here the main focus might be on the use rights, in other words, whether the household has access to draw on capital assets that it does not own (in the sense of having transfer rights to). There are differences between individuals in their ability to exercise uniform rights. Thus, the bundle of rights may be a poor proxy to explain household-level variation in access to capital. This problem has been alleviated through the concept of access, which focuses on individuals' ability to benefit from (through use and/or transfer) various forms of capital (see Ribot and Peluso, 2003).

Land is a key asset of many rural households, but its measurement raises problems. Besides the problem of defining 'land owned by the household', the valuation is challenging. Should the researcher try to value this land (and other assets) in order to construct an aggregate measure of tangible capital owned by the household? Unless there is a reasonably active land market, putting a value on land would be speculative. Land is then better measured in physical units. And rather than an aggregate measure, in many contexts one should distinguish between different types of land, for example, use, quality and location.

Differentiation should also be applied for other types of assets such as livestock: a cow is not merely a cow, but can be of different breed and age, and have different value. Disaggregation is key to get accurate estimates, although there is a balance between the accuracy and resource use. Collecting genealogy records to determine the breed of every chicken is not recommended!

Questions regarding financial, physical and natural capital are facilitated by their tangible nature, implying that theoretically coherent and quantifiable constructs can be developed to measure them. Questions regarding physical and financial assets are, however, often sensitive in the sense that people are reluctant to reveal their wealth (see Chapter 11 on 'threat factors'). Physical capital has the advantage that, in many cases, it can be observed by the researcher or enumerator, implying an easy check on whether the questions are answered correctly.

For social and human capital, the main problem is their definition and measurement. Membership of local associations is an often used proxy for social capital, but research indicates that the validity of this proxy is questionable (Bodin and Crona, 2008). For the purpose of livelihood research, a useful distinction can be made between bonding and bridging social capital, where bonding is relations of reciprocity within the community that secure a minimum subsistence livelihood and bridging is relations to more resourceful external actors (Woolcock and Narayan, 2000). Bonding social capital can be elicited through questions on who households rely on in periods of food and cash shortages or other crises, whereas bridging capital can be elicited through questions on who households rely on for assistance and support in relation to investments in productive activities.

For human capital, a widely used proxy is the level of education – both formal education and adult training – of household members. This variable may, however, not capture very well those aspects of human capital of a household that are relevant to its income earning opportunities. Locally relevant skills and experience, such as being good at carpentry, blacksmith works or producing charcoal, may be equally relevant measures of human capital.

Income

Income is the immediate outcome of livelihood strategies and a central measure of household welfare (see Chapter 5). While a well-defined concept in theory, capturing household income has proven slippery and a number of things can go wrong in questionnaire design (and during the interview).

Defining income

A consistent income definition is important to ensure construct validity (see Box 3.2). In other words, incomes from different sources should be measured consistently within the study, using standard definitions. This is also important to assure comparability of the findings to those of other studies. The definition of forest income, for example, varies across studies inhibiting comparisons (Vedeld et al, 2004).

Income is generally defined as the value added during a specific time period from assets that a household owns or has access to, such as labour, land and cattle. These assets can be used in own production and income-generating activities or sold in a market (for example, wage labour). Net transfers of cash or in-kind products are also included in income. Thus rural household income includes three broad components:

- Value added from self-employment, for example, agriculture, forestry or other business.
- Wage earnings and rents from renting out land or other forms of capital.
- Transfers, for example, remittances and pensions.

The first category is often the tricky one. The basic income equation for income from self-employment (in agriculture or business) is:

$$I = \sum_{i=i}^{n} p_i y_i - \sum_{j=1}^{m} q_j v_j$$

Income (I) is gross value (price times quantities of all n products) minus total costs (price times quantities of all m purchased inputs), for example, fertilizers, seeds, tools, hired labour.[1] In our experience, there are at least four critical issues that cause confusion and mistakes.

First, total household income is the sum of cash income and subsistence (in-kind) income, the latter referring to the value of products being produced and consumed directly by the household. Many respondents (or even researchers!) may take income to mean cash income only. It is critical that the complete definition of income (= subsistence + cash) is clear to both enumerators and respondents.

Second, the costs of household or family labour should *not* be deducted to obtain household income – this is against the definition. One may want to measure the quantity and costs of household labour for other purposes, for example, comparisons of the profitability of different activities, but it is not needed for calculating household income.

Third, some products are used as inputs in other income-generating activities, for example, fodder collected from the forest for livestock, or timber for making furniture. The researcher has two options: (a) collect data on the quantity and price of these inputs, and count the fodder as forest income, but then deduct it from the gross value of sales and consumption of livestock and livestock products; or (b) ignore it because data are too hard to get, but keep in mind that – although total income is correct – the balance between livestock and forest income is skewed. A serious mistake is to count this as forest income, but not deduct it as a cost when calculating livestock income. This will exaggerate total household income.

Fourth, the distinction between income and assets is not as clear as it may appear. Consider this example: selling a cow would, by most household surveys, be counted as income. Then, purchases of livestock must be deducted when calculating livestock income, otherwise a farmer would get higher income from simply buying and selling a cow at the same price during the survey period. But, including sales and purchases of livestock may result in some households getting low income just because they are expanding the herd size through purchases, in other words, they will be income poor and asset rich. If the income definition is to include 'changes in the value of assets', then one should also include natural growth of the livestock (and other assets). This can become quite complicated, and our recommendations for most surveys are: (a) treat assets symmetrically in the way that both sales and purchases are included for key assets such as

Box 7.2 *The importance of recollection periods*

A total income survey in Nepal included a test of recollection periods. Households were asked about environmental incomes both within one month and within three months prior to the interview. The income estimated from one month recall is consistently higher than that from three months. As the table shows, the difference between one and three months' recollection is quite high in the aggregate figures for direct forest income (unprocessed products), forest derived income (processed products) and environmental (non-forest) income. The results are in accordance with our expectations – that longer recall leads to lower income estimates because people forget their incomes. But the magnitude of the difference is perhaps surprising. Some of the difference may be because income from rare events is generalised to three months; this suggests that data collection instruments should use three-month recall periods for infrequently collected products and one-month recall for regularly collected products.

Table 7.1 Quarterly forest and environmental income figures elicited with different recall periods (Nepalese Rupees)

Income type	Recall period*	Min	Max	Median	Mean	St.dev.
Direct Forest Income	1 month	30	327,720	3300	7152	18,877
	3 months	18	112,060	2845	4808	8229
Forest Derived Income	1 month	15	790,500	1500	10,791	51,788
	3 months	10	263,500	1000	4605	17,483
Environmental Income	1 month	15	36,165	675	1565	3198
	3 months	5	37,500	300	846	2434

*Amounts from one-month's recall were multiplied by three to yield quarterly estimates.
Source: Calculation by Santosh Rayamajhi based on PEN data from Nepal.

livestock; and (b) livelihoods and poverty has both an income and asset dimension, thus modifying the income definition in an attempt to make it capture both aspects is not a good solution.

The principle of decomposition

The introduction to this chapter illustrated the need to break down the concept of income into questions that can be answered meaningfully by respondents. This is the principle of decomposition: to break aggregate values such as total household income into units that can be remembered and estimated by

Table 7.2 Example of table to record forest income

Question: What are the quantities and values of raw-material forest products the members of your household collected for both own use and sale over **the past month**?

1. Forest product (code-product)	*2. Collected by whom?*	*Collected where?*		*5. Quantity collected (7 + 8)*	*6. Unit*	*7. Own use (incl. gifts)*	*8. Sold (incl. barter)*	*9. Price per unit*	*10. Type of market* (code-market)	*11. Gross value (5*9)*	*12. Transport/ marketing costs (total)*	*13. Purch. inputs & hired labour*	*14. Income (11-12-13)*
		3. Land type (code-land)	*4. Ownership* (code- tenure)										

Source: PEN (2007)

Note: the four general codes used (plus one specific for this table to column 2). Also note the logical build up in order to calculate income.

respondents. Whereas a respondent is unlikely to know what the total income from agriculture was in the just-ended season, she may, if asked questions about the quantities of each crop harvested and the prices at which they were sold, be able to give quite accurate responses. Thus disaggregating income into prices and quantities is a first step in decomposing or disaggregating income. Other common ways of further disaggregation include asking questions separately by:

- Product, for example, crops for agricultural income.
- Land area, for example, plot for agricultural income, and forest area for forest income.
- Household member, for example, for wage income.
- Season, for all incomes that show seasonal variation.

When using disaggregation it is, of course, important to make sure that the disaggregated units add up, for example, that the right income definitions are used. Decomposition, however, risks violating the principle of construct validity of the income estimate through omissions or double-counting of income components. If the data are broken down to the needed detail, double-counting can be corrected for in the analysis phase. Still, the general recommendation is to make the questionnaire design reflect correct income definitions.

Recall periods

While we clearly recommend a product-by-product approach to obtain sector income, the question of seasonality is a major headache. Most studies seek to obtain an estimate of annual income. The preferred option is to do regular surveys over one year, as done in PEN with four quarterly income surveys. The recall period is short (three months or less), and seasonal variation is captured. But often researchers do not have the resources for repeated visits, and therefore end up doing one-shot surveys. The researcher then has three options:

1. Ask about income for the last 12 months (appropriately decomposed, for example, by product).
2. Ask about income for, say, the last month or last three months, and multiply to get the annual income.
3. Divide the year into a few (normally two or three) distinct seasons, and ask about income in each of these.

The strategy depends on the income type and the local context. In many countries there are two main agricultural seasons. In some countries there are three. A survey that tries to estimate annual agricultural income should have separate tables/sections for each season. In Tanzania, these would correspond to the *mvua* (rainy) and *kiangazi* (dry) seasons. In rice-producing areas of Bangladesh, they would refer to the *aman*, *aus* and *boro* harvests. Similarly, collection of firewood can vary by season, thus it is relevant to decompose data collection into, for example, the rainy and the dry seasons. If repeated surveys are not possible (always the first best choice), we will – as a general rule – recommend option 3 of seasonally specific questions. Exceptions should be made for incomes that do not display much seasonal variation.

The PEN surveys had a recall period of one or three months for the quarterly income surveys (Chapter 1). Small and frequent activities, such as collection of firewood for household consumption, favour short recall. Infrequent activities – be they small, such as collection of mushrooms in the rainy season for own consumption, or large, such as purchases of land or livestock – require longer recall to be sure to capture the income. Unfortunately, the choice of recall periods can have significant impacts on the resulting income estimates. This is illustrated in Box 7.2 from an experimental study in Nepal.

The ' + ' in the HAI +

Finally, the '+' in the HAI+ denotes all other data needed to answer the specific research questions. Here we discuss just some of the data that might be relevant to collect. For example, in research on livelihoods it may also be of interest to gather data on consumption expenditures and/or household labour allocation. These types of data allow for interesting livelihood analyses, but collecting them is also notoriously time-consuming.

Like income, getting reliable consumption data requires very careful recording of all consumption items. This is particularly true if you want to use aggregate consumption as a measure of household welfare; collecting data on just a few consumption items will not provide the full picture. Moreover, the recall period should be short for consumption expenditures (a few weeks, and not years). But it will also depend on the frequency of purchases: for rice it can be the purchase last month, while for bicycles it should be last year.
Some simple questions regarding consumption can be useful in order to check the validity of the income data. In case the enumerators suspect some income sources to be forgotten or under-reported, it can be revealed by simple calculations based on questions such as: 'How often do you go to town to make purchases? How much on average do you have to spend?' Depending on the answers, follow-up questions could be asked.

Household labour allocation data are useful for a number of purposes, for example, to calculate the return to labour in different activities the household engages in. But, it takes a lot of time to get reliable data (ask for each household member, for a number of tasks, and have short recall periods). Also remember that household labour data are not needed to calculate household income.

Often one is interested in particular activities that the household members engaged in: a microcredit scheme, a forest user group, a women's association or a marketing association. The questionnaire could then include questions related to membership, why they are (not) members, perceived and/or concrete benefits and costs of participation, how active they are in terms of time spent, and so on. Such information is often useful both to provide a general background and for more direct statistical analyses, for example, to explore differences between rich and poor households in their motivation for participating in a particular programme.

Collection of data on values and attitudes are often included in household surveys, making the issue of construct validity increasingly challenging. How does one, for example, ask about the level of transparency in the allocation of grazing rights to village common pastures? If possible, such data needs should therefore be met through other data collection methods (see Chapter 6). If eliciting values and opinions through a survey cannot be avoided, the Q-sort methodology provides a widely applied approach (Gray, 2009). A Q-sort approach to elicit values and opinions would ask respondents to rank-order a set of statements about a topic on the basis of their individual points of view. By this, respondents reveal their subjective viewpoint or personal profile (van Exel and de Graaf, 2005). Using the Q-sort methodology provides a structured approach to overcome issues regarding question form, wording and context that have been shown to affect answers dramatically, in particular when eliciting values and opinions (see Schuman and Presser (1996) for a treatment of these issues).

Formulating the questions

There is plenty of evidence that even very simple questions are answered incorrectly. For example, one study found that 10 per cent of respondents in a Philadelphia, US, survey gave different answers to the question, 'What is your age in years?' when re-surveyed a week later (Foddy, 1993). Given this, one must wonder about the validity of answers to more subtle or complex questions.

Answers to questions are affected by the question format. Foddy (1993) reports on a study in Australia that found that for a particular magazine, only 7 per cent of respondents said they bought it if asked using an open question ('which magazines do you buy?') while 38 per cent did so for a closed question ('which of the following magazines do you buy?'). In the context of surveys, writing the question so that enumerators can read them aloud directly is one way of increasing the likelihood that the same questions are posed similarly by different enumerators. It may also be necessary to specify allowed and sequenced lists of probes to avoid bias (see also Chapter 11).

In relation to more tangible issues, such as assets and incomes, the precision of the question becomes an issue in relation to wording, as reflected in the example in the introduction to this chapter. It is important to leave as little space for individual interpretation of the question as possible to minimize this source of (generally unwanted) variation.

Characteristics of good questions

Based on our experience, here are some of the characteristics of good questions.

- **KISS**: Rule number one is KISS: keep it sensibly simple.
- **Concrete and specific**: Be concrete, and avoid hypothetical questions. For behavioural studies, collect data on revealed choices rather than expressed preferences for hypothetical choices. For example, do not ask: 'how would you respond to a food shortage?' Instead, ask if the household has experienced a serious food shortage in the last three years and what was done in response to the shortage.
- **Short recall**: Questions should in general not require long-term, detailed memory. The rule should be, to reformulate Einstein: 'Keep the recall period as short as possible, but not shorter.'
- **Local units**: Questions should be formulated to allow respondents to report measures in locally understood units (for example, in scotch carts, buckets, bags) rather than in unfamiliar metric units. A survey to standardize units can then be conducted later/concurrently to establish conversion factors for the local units into more standard measures if required to answer the research question.
- **Quantify answers:** If possible, answers should reveal quantities. Consider the question: 'How often do you go fishing?' A poor coding system would use categories such as: very often, often, sometimes, and so on. A better coding system is: more than once a week, 1–4 times per month, and so on. And, an even better formulation would be to ask 'how many times per month do you go fishing?'. Even this formulation is not perfect for two reasons: First, be careful with the term 'you' if the meaning is 'any member of the household'. Second, the fishing pattern may vary over the year. Thus, a good formulation taking all these factors into account would be: 'How many times over the past 30 days did any member of your household go fishing?' (Although this formulation does not factor in the variation in fishing

patterns over the year.) If recall is difficult, it may be necessary to help the respondent by asking about fishing during the last week followed by asking if this week was typical for the last month.

- **Define terms carefully**: Be careful with terms used. The question 'how much agricultural land do you have?' may or may not include land rented in/out, and may or may not include land that is currently not under cultivation (fallow land). Also, if asking the husband, he might only think of land that he is in charge of, and not land cultivated by the wife.
- **Avoid multipart questions**: Ask one question at the time, and avoid multipart questions. For example, do not ask: 'Have you ever had or applied for a microcredit loan?' It is better to first ask if they have had a loan, and – if no – ask if they have applied for and been rejected a loan.
- **Neutral formulations**: Many respondents will try to 'please' the enumerator and researchers, particularly if asking about preferences, perceptions or behaviour. Even a seemingly simple question such as 'would you like to get a microcredit loan?' may be interpreted as getting a loan being something positive. A better formulation might be (although it might be too hypothetical in some contexts): 'If you today were offered a microcredit loan of x shillings, to be paid back over y years at an interest rate of z, would you accept the offer?'

Non-factual questions

Questions related to preference, perception and behaviour typically have a different format than more factual questions, such as 'how many bags of rice did you produce last harvesting season?'. Consider microcredit again – this time it is necessary to find out why some respondents do not have a microcredit loan. The question might be: 'Why have you not had any microcredit loan?' Three main classes of reasons can be expected: (a) there is no microcredit institution offering loan in the village (supply), (b) there are but the respondent is not interested (demand), or (c) microcredit loans exist, but the respondent is not informed about it (information). To find out which category a respondent belongs to, the first question can be if microcredit is available in the village. If the answer is affirmative the questions follows as to why the respondent does not have a loan. There are then several options for asking the questions, including:

1. **Open-ended, no coding of responses**: With open-ended questions one simply writes down the responses. Although these can be coded during data entry, they are often difficult to deal with when analysing data and in many cases end up being abandoned. Open-ended questions are better in village questionnaires, focus group or key informant interviews. They are also

useful during the initial exploratory stages of research, to learn about the area and prepare a systematic questionnaire (see also Chapter 5 – contextual information).

2. **Open-ended, responses coded**: An alternative is to ask an open-ended question, with a predefined set of possible answers that are not read to respondents (including 'Other').
3. **Closed-ended**: Ask the question and read out the different (pre-coded) alternatives for responses.

Respondents may have more than one reason for not having had a loan in the example above, and with options 2 and 3 for asking questions one of the following three formats for recording the response may be chosen:

1. Tick the responses that apply (no ranking).
2. Give a rank 1–3 (or 5) for each possible response, depending on importance (very important – important – not important).
3. Rank the responses in terms of importance from 1 to 3.

In PEN, we opted for asking open-ended questions, with responses coded and filled in during the interview (option 2). If the options are presented verbally, respondents tend to go for the last one (this is a common example of response-order effects; see Schuman and Presser (1996) for a thorough treatment). We also ranked the responses (maximum of three), that required a follow-up question, once the different reasons had been listed on their relative importance. This was done to get information on the importance of the different alternatives. A combination of question option 3 and recording option 1 (read all responses and tick) risks yielding too many responses and failing to grasp what really matters.

Questionnaire layout

The layout of the questionnaire is concerned with the presentation of the questions to the enumerators and, subsequently, respondents. Generally speaking, the main consideration in the design of the questionnaire layout is to minimize the risk of problems when the enumerators present the questions to the respondents.

Every questionnaire should have a fully spelled out introductory paragraph that introduces the research to the respondent and asks permission to actually conduct the survey interview. Such an introduction should have the following components (Rea and Parker, 1997):

Box 7.3 *Some practical tips for questionnaire design*

- Many questions are better drafted into tables for ease of recording, coding and data entry (Table 7.2). This assumes that you have good enumerators, thus no detailed 'word for word' question formulations are needed for all questions.
- Assign an identification number to each household member to assure that each person gets a unique identification. This can be used in several sections of the questionnaire (for example, household composition, wage income and participation in savings groups).
- Pre-code as many responses as possible but leave room for other responses. Check and refine the codes during pretesting.
- Common codes can be included in a separate code list (for example, product codes) to save space.
- Leave enough space for enumerators to be able to take notes and provide explanations.
- Write key instructions for enumerators (and use, for example, *italics* to separate from text to be read out to the respondents).
- Spiral binding of (sets of) questionnaires helps in getting organized and avoiding loss of some questionnaires or questionnaire pages.
- Use major section codes for each section (for quick reference).
- Page numbers on each page of the questionnaire.
- Assign a unique code to each enumerator in case of problems later.

- The organization conducting the study should be presented. This is important to allow respondents to know who the enumerator represents and who asks the questions.
- The objectives and goals of the survey should be stated. It may be useful to explicitly state how the objectives relate to the respondents to generate as much interest as possible on behalf of the respondents.
- The basis of the sample selection should be made clear to allow the respondent to know how and why she was chosen for the research.
- It must be assured to the respondent that her participation is valued and that all information surrendered will be treated with confidentiality.
- An estimate of the duration of the interview should be stated.
- Permission to conduct the interview should be asked before the enumerator can proceed to the actual research questions.

Furthermore, every questionnaire should have a section with process information, in other words, information concerning who did the interview, checking of interview data, coding, data entry and data entry checking, and when these were done. This is valuable information for tracking down systematic errors caused by human agency.

The ordering of the questions and sections is important for several reasons. Avoid putting off respondents by having too many boring start-up questions (Gray, 2009). Asking about the H-part, the family members, can often be a soft start, and shows an interest in family life. The most important sections should also be taken early. Sensitive questions (regarding assets, debt, illegal activities, and so on) should be kept at the end. Answers to earlier questions can affect answers to later questions. If you ask Danes to rate how 'Danish' potatoes are, they rate them as more Danish if you first asked them how Danish rice is. Earlier questions can either reinforce or work against the response given.

A few practical tips on questionnaire layout are given in Box 7.3.

Conclusions

Designing a good questionnaire is both a science (clear rules to follow) and an art (skills based on experience, see Box 7.4). There are some useful general rules and experiences to draw on regarding question sequencing and formulation, but nothing can replace pretesting (see Chapter 10) and critical review by fellow researchers.

The chapter has highlighted some of the issues to be aware of when designing the survey questionnaire and formulating questions. Overall data needs need to be decomposed into questions that can be posed to respondents, without compromising construct validity in the process. One of the largest threats to construct validity of research based on questionnaire surveys is incorrect 'translation' of research questions and hypotheses into questions that go into the questionnaire. Also, the process of decomposition often runs the risk that important data needs are not reflected in the final questionnaire or that unnecessary questions are added. Even with correct 'translation' of research questions and hypotheses into questionnaire questions, poor data quality may be the result of poorly formulated questions that respondents are unable to answer.

The wording of the questions that go into a questionnaire need to follow some overall guidelines: be concrete, specific, simple and use neutral formulations and carefully defined terms. Finally, overloading the questionnaire should be avoided as it comes with high costs and an added risk to data quality

Box 7.4 *Learning from the Living Standards Measurement Study*

The Living Standards Measurement Study (LSMS) was established by the World Bank in 1980, with an aim to increase the accuracy and policy relevance of household survey data collected in developing countries. The programme was designed to identify how policies could be designed to positively affect social outcomes in health, education and economic sectors, and so on. Since the first LSMS survey in 1985, it has been implemented in more than 30 countries with repetitions in many of these.

A typical LSMS survey gathers detailed information at household level regarding household composition and demographics, education, health, employment, migration, housing, consumption, agriculture, enterprises, income, savings and credit. Some of the information is collected at the level of the individual household member. In addition, the LSMS gathers information at the community level and also typically includes questionnaires focusing on local prices.

All the experience gained from the LSMS project is available through the LSMS Working Paper Series that can be found on the homepage of the World Bank.[2] At the time of writing, there were 135 such reports. The three volumes by Grosh and Glewwe (2000) provide detailed guidance on the construction of multi-topic survey questionnaires and question formulation. Overall points from the LSMS project are:

- The starting point for designing modules and questionnaires of a multi-topic survey is a set of policy questions.
- When designing multi-topic surveys involvement of the right people in the process is a prerequisite for success.
- Designing a multi-topic survey involves a host of trade-offs.
- Collecting comprehensive household survey data in developing countries is feasible.
- Two main problems have been to measure (a) household income from agriculture and non-agricultural self-employment, and (b) savings and financial assets.
- Few LSMS surveys have collected data to examine environmental issues.

Source: Grosh and Glewwe (2000)

because of respondent fatigue. The process of questionnaire development often involves 'shaving' the questionnaire. The aim should be to bring the average interview time (not testing as it will take longer) down to a maximum of one hour.

Key messages

- Ensure that research questions and hypotheses are translated into questions that can be meaningfully answered by respondents without losing construct validity.
- The wording of the questions that go into a questionnaire should be concrete, specific, simple and use neutral formulations and carefully defined terms.
- Use the principle of decomposition to collect data on household income (disaggregate by product, land area, household member and season).
- When checking the question formulation, try actively to misunderstand each question.

Notes

1 Maintenance of capital stock (or depreciations) should also be included, but this will have limited applicability for most households.

2 http://econ.worldbank.org/WBSITE/EXTERNAL/EXTDEC/EXTRESEARCH/EXTLSMS/0,,contentMDK:21555770~menuPK:4196843~pagePK:64168445~piPK:64168309~theSitePK:3358997~isCURL:Y,00.html

References

Barzel, Y. (1989) *Economic Analysis of Property Rights*, Cambridge University Press, Cambridge

Bodin, Ö. and Crona, B. I. (2008) 'Management of natural resources at the community level: Exploring the role of social capital and leadership in a rural fishing community', *World Development*, vol 36, no 12, pp2763–2779

Bromley, D. W. (1991) *Environment and Economy: Property Rights and Public Policy*, Blackwell, Oxford and Cambridge

Cavendish, W. (2002) 'Quantitative methods for estimating the economic value of resource use to rural households', in Campbell, B. M. and Luckert, M. K. (eds) *Uncovering the Hidden Harvest: Valuation Methods for Woodland and Forest Products*, Earthscan, London, pp17–66

de Long, S. J. (1997) *Regression Models for Categorical and Limited Dependent Variables*, Sage, Thousand Oaks, CA

Foddy, W. (1993) *Constructing Questions for Interviews and Questionnaires: Theory and Practice in Social Research*, Cambridge University Press, Cambridge

Gray, D. E. (2009) *Doing Research in the Real World*, Sage, London

Grosh, M. and Glewwe, P. (2000) *Designing Household Survey Questionnaires for Developing Countries: Lessons from 15 Years of the Living Standards Measurement*

Study, vols 1–3, World Bank, Washington, DC, http://econ.worldbank.org/WBSITE/EXTERNAL/EXTDEC/EXTRESEARCH/EXTLSMS/0,,contentMDK:21556161~pagePK:64168445~piPK:64168309~theSitePK:3358997,00.html, last accessed 5 February 2011

PEN (Poverty Environment Network) (2007) *PEN Technical Guidelines*, Centre for International Forestry Research, www.cifor.cgiar.org/pen/_ref/tools/guidelines.htm, last accessed 5 February 2011

Rea, L. M. and Parker, R. A. (1997) *Designing and Conducting Survey Research: A Comprehensive Guide*, Jossey-Bass Inc, San Francisco, CA

Ribot, J. C. and Peluso, N. L. (2003) 'A theory of access', *Rural Sociology*, vol 68, no 2, pp153–181

Schuman, H. and Presser, S. (1996) *Questions and Answers in Attitude Surveys: Experiments on Question Form, Wording and Context*, Sage, London

van Exel, N. J. A. and de Graaf, G. (2005) *Q Methodology: A Sneak Preview*, www.qmethod.org/articles/vanExel.pdf, accessed 1 September 2010

Vedeld, P., Angelsen, A., Sjaastad, E. and Berg, G. K. (2004) 'Counting on the environment: Forest income and the rural poor', *Environmental Economics Series*, no 98, World Bank, Washington, DC

Woolcock, M. and Narayan, D. (2000) 'Social capital: Implications for development theory, research, and policy', *The World Bank Research Observer*, vol 15, no 2, pp225–249

Chapter 8

Valuing the Priceless: What Are Non-Marketed Products Worth?

Sven Wunder, Marty Luckert and Carsten Smith-Hall

The cynic knows the price of everything and the value of nothing.
Oscar Wilde (1893, *Lady Windermere's Fan, Act 3)*

Introduction

Households in developing countries collect and use a wide range of environmental products, from foods and construction materials to medicines and composted manure. In many remote rural areas, the bulk of goods collected by households, rather than being sold, is destined for direct consumption (for example, subsistence consumption of game or construction poles) or used as inputs into domestic production processes (for example, fodder for livestock). In these cases, there is no explicit transaction price that we can use for valuing the quantities of goods consumed.

Still, we may want to value these non-marketed activities for several reasons. First, we could be interested in a measure of the welfare contribution that different types of natural resources provide to households, that is, to estimate environmental income. Second, these values may help us understand how and why households allocate their labour, land and capital across different income-generating activities. Third, values of natural resources can also be important for policy-makers. For instance, if policies promote forest clearing for agricultural expansion, how much forest-derived non-marketed income is lost in the process? Fourth, in poverty analysis we need to get a value estimate of overall household income and consumption.

Values are thus the basis of any analysis of households' livelihoods. Counting physical quantities of products does not tell us much about their contribution to well-being. Instead, we seek to enumerate these resources in terms that provide us with insights into the welfare implications and decisions, at household and policy-making levels, of how resources are used. But expanding enumeration from physical quantities to values is fraught with difficulties. In this chapter we provide an introduction to some key concepts and methods. In the following section, we take a closer look at the value concept. Next, we describe some of the structural obstacles found in peasant economies of developing countries, which frame the resource valuation problems that we want to address. Keeping these features in mind, we then outline and review six different practical methods for how to assign values to non-marketed goods. The last section gives some suggestions regarding how price data can be checked to see if they are reasonable.

What is value?

Values may be thought of as measures of how much people want or like various goods and services. The concept of values arises from the belief that there exists a common expression of benefits to people that can be expressed and aggregated across numerous types of resources and individuals. As such, we may use these aggregated values as expressions of well-being of people who hold diverse livelihood portfolios.

The concept of values can encompass broad concepts of what people like. One distinction is between 'held' and 'assigned' values (Adamowicz et al, 1998). When we observe a market price, we are seeing an assigned value that is thought to be transient, changing with market conditions. Assigned values may be conditioned by held values that are more basic, more qualitative and based on morals and preferences that are thought to change more slowly, if at all. In the following, our exclusive interest will be in assigned values. However, the diversity of people's underlying broader value systems should always be kept in mind. For instance, people may value certain forests, mountains or rivers for their cultural or religious purposes, and often (legitimately so) be unwilling to translate these values into monetary figures. The quantitative values we assign are thus bound to be incomplete measures of the multidimensional sources of human welfare.

The quantitative values we assign to resources depend on the alternative scenarios we imagine for them. For instance, my *utilitarian value* of an asset (say, a horse cart) represents the welfare loss I would experience without that asset, for example, having to carry things by hand instead. The *production value* reflects the inputs of labour and capital that were needed for making the cart. The *sales value* is the money I would receive by selling the cart in the market. The *replacement value* becomes relevant had I to substitute it for a similar asset, for example, a handcart instead of a horse cart.

Values are frequently exchanged in markets and expressed as prices that buyers pay to sellers. Sellers are thought to focus on costs of alternative forms of production, while buyers are thought to focus on benefits derived from alternative possible purchase combinations. The considerations of sellers and buyers come together in markets, which specify prices at which trades of goods and services occur. The higher the market price of a product, the more providers will be induced to sell it and the fewer buyers will demand it. Hence, prices change over time in response to changes in preferences of buyers and/or costs to sellers.

Although markets may frequently produce a single price for a given product, it is not reflective of the value of a good or service to all people. There is frequently great heterogeneity among individual sellers and buyers. Thus, many buyers 'find good deals' by buying products at prices that are less than the maximum that they would be willing to pay. The difference between the actual price paid and the maximum willingness to pay is referred to by economists as *consumer's surplus*, and expresses incremental values that buyers receive from market transactions. Similarly, normally many producers sell their products for more than the minimum amount that they would be willing to accept. Such differences are referred to as *producer's surplus,* and are taken as a measure of incremental benefits that sellers derive.

Prices arise in response to specific supply and demand conditions, which respond to their scarcity at the margin but will not necessarily mirror the innate usefulness of resources in absolute terms. For example, Box 8.1 indicates that low firewood prices do not mean that firewood is not valuable to households. Rather, low prices are a reflection of the abundance of firewood that reduces the costs of supply. If firewood was less abundant, we would expect to see higher prices.

Because the value of a product is so closely tied to its abundance, it is important to carefully consider the quantity of resources that are being valued. For example, if valuing the change in firewood consumption of an individual household, it may be reasonable to assume that the individual household is not largely affecting the abundance of firewood. Economists frequently take this 'marginalist' approach in that they value things based on the relative scarcity of goods and assets vis-à-vis small counterfactual changes. In such cases, economists may seem to 'take nature for granted', in other words, assign low values to natural resources that are abundantly available. In many cases, such as in Box 8.1, this approach may yield an accurate reflection of values. But if the true counterfactual is a large-scale, devastating deterioration of natural assets (such as climate change) then it becomes more difficult to estimate values marginally based on counterfactuals constructed from current conditions. For example, Costanza et al (1997) attempted to put values on entire global ecosystem functions, assuming these would have to be fully lost or replaced. While these estimates might portray nature's worth in high-level policy arenas, the estimates are difficult to defend because counterfactuals are not clearly defined.

Box 8.1 *Bringing Adam Smith to the field*

When Adam Smith published his famous *The Wealth of Nations* in 1776, he used the so-called 'water-diamond paradox' to didactically illustrate the power of marginality in determining economic value. Water is essential for all life on Earth, yet since it is usually in abundant supply, it normally cannot be sold – at least, at the time of his writing. In turn, diamonds are a luxury commodity of limited direct use value – life could easily go on without them – yet they are highly priced for being scarce. Supply and demand thus determine exchange values, sometimes in contradiction to the logic of use values.

About 230 years later, Manyewe Mutamba conducted a village interview in Mufulira District in the Zambian Copper Belt, asking villagers which forest product was the most important to them. But, 'most important', for what: food, shelter or cash? Economists typically rank between 'apples and oranges' by assigning prices to different commodities, but in this case, most products were for subsistence use only. There was no intuitive reductionist yardstick. He thus asked people: 'Which one would be the product that it would be most difficult for you to lose?' Surely that clever hypothetical question would force them to prioritize! After some internal discussion, the group consensus was: 'Firewood.' Why? Because without firewood, it would be impossible to cook – and that would clearly be a major disaster.

For an economist, this was a surprising response: a forest walk revealed that wood resources remained extremely abundant. Since there was no shortage of firewood, it would also not have any mentionable exchange price and thus be of little economic value. Moreover, any intervention producing marginal changes in firewood availability would also have negligible influence on peoples' livelihoods. So, why bother about firewood?

The seemingly perplexing answer was fully explained by the nature of the question. We had not asked people about what commodity they would be most worried to lose *at the margin*, but in totality. We had asked them about which product had the highest use value, not exchange value. The scenario we implicitly had given to them – the prospect of losing all access to firewood – was a counterfactual completely outside of their local reality, without relevance in any foreseeable future. Firewood to them was what water was to Adam Smith.

Prices are not only employed as measures for individual households' welfare. We also frequently use prices as indicators to optimize welfare and resource allocation at higher aggregation levels, such as villages, districts or nations. For instance, if a district government was to decide whether to establish an agro-industrial project, it should evaluate the costs and benefits by also looking at local prices. This would refer to the prices of the incremental agricultural products that will be produced, but perhaps also of currently extracted forest products, which may partially be lost if the project implied conversion of forest to cropland. In this type of cases, prices generally most accurately reflect social values when:

- markets are competitive (in other words, there are sufficient buyers and sellers so that neither can individually influence prices);
- there are no 'market failures' with external side effects of production and consumption activities on third parties (or, alternatively, such failures are being corrected for in separate markets, for example, for environmental services); and
- markets create distributions of income that are in line with social desires to promote equity.

But markets may fail on any or all of these counts. For example, only a few dominating wholesale intermediaries might offer artificially low prices when buying in rural markets, and secure artificially high prices when reselling in urban centres. Production costs of gold panning may factor in costs of back-breaking labour necessary to find nuggets, but neglect environmental costs of streamside erosion from mining practices. In harvesting firewood at low prices, households may fail to leave woodlands for future generations. Despite concerns in this regard, they may fear that individual restraint would make no difference because other households would take over their share of an open-access resource (in other words, property rights matter).

Finally, markets may create inequities among village members that are unacceptable to people's concepts of justice. All of these scenarios create situations where prices fail to accurately reflect social values and where the analyst needs to make adjustments. The issue of how to do social valuation and cost – benefit analysis, however, goes beyond our purposes here; for a description of these issues with an environmental angle, see Hanley and Barbier (2009).

Rural livelihoods and prices

People living in developed economies typically produce products and services they sell for a living and for their monetary receipts (salaries, profits and transfer incomes), they buy their necessities: production and consumption decisions are clearly separated. They can also normally buy and resell (or vice versa) the same product – for example, a used car – at a reasonable price margin: the loss from reverting most trades is manageable. Typically people have access to credit markets to generate some liquidity when needed, at reasonable costs. And through insurances they can safeguard themselves against major risks.

All of these circumstances can be quite different in rural areas in developing countries, where market imperfections tend to be much more pronounced. Often there are no insurance mechanisms, and credit is perhaps only available at usury interest rates. People typically produce some goods just for sale, some for subsistence use and many mixed for both purposes, such as selling in the market the surplus of staple food production, once basic household needs have been met. However, margins between buying and selling prices tend to be much larger: middleman profits can be high and risks of price fluctuations can be high if markets are thin (in other words, low trading volumes make prices jump frequently) or seasonal (for example, before and after peak harvests). Most of all, the transport costs of getting commodities to and from the market can be very high. This means that rural producers in developing countries may face relatively wide price bands between selling and buying prices of a product (Sadoulet and de Janvry, 1995).

What does that mean for our valuation and pricing problem? Let us have a look at an example. Let us say that, as a farmer at a tropical forest margin, I am producing maize. The market price in the nearest town is 20 shillings per sack, but the high transport costs on a dirt road, and the intermediary profits taken by a single transporter monopolizing the trade, sum up to 7 shillings, each transport way. Hence, in the village, the selling price I am being offered by the middleman is just 13 shillings. In case I occasionally needed to buy maize, the buying price of maize is 27 shillings. This leaves an external margin, or a price band, of $7 + 7 = 14$ shillings. In this price band, going from the price levels of 13 to 27 shillings, trading is not favourable to me – unless when I am occasionally trading maize with my neighbours within the village, with much less transaction costs and on more equal terms. The price I would be willing to sell maize for is 15 shillings, which is also called my 'shadow price' – in other words, an invisible price where economically things would break even for me, making neither losses nor gains. So, if hypothetically I could access the urban market at no transaction cost, I could sell maize at a competitive price of 20 shillings and make 5 shillings of profit per sack. But due to the elevated transport and commercialization costs, I can only sell at 13 shillings, and would thus actually lose 2 shillings.

Let us now say that I could make my maize production more efficient, and produce at 10 shillings instead of 15 shillings per bag. That would allow me to make 3 shillings of profit by selling to the middleman (13 minus 10 shillings). However, there might still be reasons for me to hesitate with this deal. If my production is just a bit higher than my household consumption, the next harvest in the village could be a bad one, so that I would have to buy maize from outside to feed my family. If the buying price is a prohibitive 27 shillings, that implies a huge risk – especially if I cannot get access to credit or insurance. It might thus be better for me to hedge against future risks by saving the extra sacks of maize for a rainy day – unless they would likely perish during storage.

The large non-traded price band here makes it unattractive for me to participate in the maize market. If my neighbours are in a situation with similarly ranged shadow prices, the potentially tradable product, maize, would thus in our village become a de facto non-traded good, due to the high transaction costs involved in trading. Note that my price band problem could be even worse for more bulky products such as construction poles or firewood, because the transport costs here make up a higher portion of the final market value than for maize, or for perishable products such as fruits or vegetables, where a larger share of the products may become physically lost or damaged during the transport and commercialization process.

Along comes now to the village a young PhD student who wants to measure my household's welfare, and how much value different livelihood components contribute to it. We are discussing what price to value my maize production with: clearly both the intermediary's selling and buying prices would be inadequate, as would in this case also be the intermediate urban market price. The value we are looking for is my household's shadow price, which is jointly determined by my production costs and preferences for own consumption of the product in question. But that price is not stated anywhere, and my own gut feeling about its size might not be precise enough. What options exist for the young researcher to get a good price proxy? The next section outlines some hands-on ideas.

What valuation methods to use?

The choice of valuation method should generally be tailored to the specific characteristics, including the objectives of the study, the presumed importance of different types of goods and services, and the local information that is available. Our focus here is on the private benefits enjoyed by households, for subsistence and sale. Other environmental benefits include ecosystem services (for example, protection of watersheds, biodiversity, carbon stocks and recreational values) that can be valued through a series of methods (for example, hedonic pricing, travel costs, defensive expenditures and replacement costs, production function approaches), which will not be treated here; see IIED (2003) for a general description of forest-benefit valuation techniques in developing countries. For quantifying household benefits in particular, we recommend the following six methods:

Local-level prices

Whenever available, using local prices is the first choice. These prices can come from within-village transactions or farm-gate/forest stumpage prices, and are extrapolated as general value indicators even to people who consumed but did not trade the product. Using the example from the previous section, if there is informal trade of maize between households within the village, the price used in these transactions might be a good proxy for valuing maize, in the absence of any external trades. It may be necessary to use focus groups and/or small market surveys to obtain the desired information.

Discussion: The big advantage here is that there exists a local price revealed by a real-world transaction. However, care is needed with this method when the underlying markets or transactions are extremely thin and unrepresentative. For instance, say maize is only being traded seasonally right before the harvest, when scarcity is at its peak and prices are thus very high. Or let us assume that bushmeat is consumed by everybody in the village, but it is only being bought by the wealthiest households, who engage in more rewarding activities that do not leave them enough time to go on long hunting trips and they thus likely have a larger willingness to pay for bushmeat than other people in the village. Both of these features would make reported prices too high for extrapolating to the desired value of common consumption over the entire year and population.

On the other hand, when I am buying maize from my neighbour, he might charge me a lower-than-normal amount, because he wants me to help him in the construction of his new stable: there is an expectation of a return favour embedded in the low price, which thus also constitutes an investment in social networks (Rao, 2001). Similarly, the farm-gate price that forest extractors of rubber or Brazil nuts receive from an intermediary trader is often low, because the trader has provided credit in advance to the extractors. In both of the latter cases, the local price would underestimate values, because it is invisibly bundled with other benefits.

Barter values

A non-traded commodity may locally be bartered for a marketed commodity. For instance, assume mushrooms are not traded, but occasionally exchanged between households for rice, which is usually a highly traded staple. Hence, rice can serve as our *numeraire*: a common value measure that through triangulation implicitly sets a price for mushrooms: if 1kg of rice is commonly exchanged for one bag of mushrooms collected in the forest, and the former costs 20 shillings, then this price is also valid for the latter non-traded product.

Discussion: Barter values are as good as direct trade in reflecting de facto values, and as such an attractive measure – if one can find them. Barters may in many economic contexts have ceased to exist, giving way to cash transactions. Barters may perhaps even more than cash transactions be influenced by the aforementioned 'return favours' from social relations that are embedded into inter-household transactions, thus underestimating values. If variation between the implicit prices contained in different bartering deals is high, we should be suspicious (see also data checking in next section). We could also use *hypothetical* barters to elicit a proxy for a market price ('how much rice would you be willing to accept for your mushrooms?'). This is a contingent valuation approach (see next point), just with a non-cash *numeraire* being used.

Contingent valuation

In the absence of any monetary or barter transactions whatsoever, one can ask respondents directly about their hypothetical maximum willingness to pay (WTP) or minimum willingness to accept (WTA) for a non-traded item. The choice between WTA and WTP should be determined by the most likely counterfactual – in other words, whether an item is locally most likely to be bought or sold – but normally WTP is more reliable. Contingent valuation is a common and consolidated stated preference method in environmental economics, which has enjoyed increasing popularity, especially for valuing public goods (Mitchell and Carson, 1989; Hanley and Barbier, 2009). One case study in Ethiopia used contingent valuation (WTP) to elicit the benefits perceived by villagers from a community forestry programme – both for public and for private non-marketed benefits (Mekonnen, 2000).

Discussion: While contingent methods have become fairly standard in developed countries, in developing-country settings, more so in rural areas, respondents may culturally have much greater difficulties answering contingent questions that attempt to put monetary values on non-traded items (Whittington, 1998). Alternatively, they may answer strategically, in other words, understating or overstating values they suspect might influence posterior interventions by donors or lawmakers (see also Chapters 10 and 11). A second critique of contingent methods is more fundamental: it measures preferences of the individual being questioned, which, unlike in a marketplace, includes not only a (hypothetical) market price, but also the individual's consumer surplus – in other words, what we called above the 'bargain hunting' of obtaining goods cheaper than the utility derived from them. This means that large differences between individual WTPs may occur, which as a proxy of aggregate market values lead to an overestimation of values.

How can contingent methods be used in rural household surveys? If valuation of a single received benefit (for example, a public good) or, conversely, the opportunity cost of giving up one (for example, those from avoided deforestation) is the primary focus of the research, then great care is needed in formulating the hypothetical questions in accordance with the theory of, and accumulated experience with, contingent analysis. Even so, in remote rural regions, one should expect only mixed chances of success with this method. If the purpose is more pragmatic, such as to receive value range estimates for non-traded subsistence products, the task might be easier. In the Poverty Environment Network (PEN) project, several scholars used the hypothetical WTP in focus groups to collectively value certain subsistence products, thus obtaining a consensus estimate. This collective consolidation could also help in reducing the aforementioned consumer surplus bias in individual WTP estimates.

Substitute goods values

Marketable close substitute goods might help providing useful value approximations. These can be either similar goods (for example, using a marketable timber species for a non-marketed one) or an alternative good (for example, a pharmaceutical product instead of medicinal plants). As for the second type, locally non-traded firewood is often being valued by comparing its energy content with commercial local close substitutes, such as gas, kerosene or soft coke – the latter being used, for example, by Chopra (1993) in an effort to value Indian tropical deciduous forests. Similarly, Adger et al (1995) and Gunatilake et al (1993) use substitute pricing for obtaining non-timber forest product (NTFP) values for building materials, medicines, firewood and fruits in a national forest valuation for Mexico and valuation of local uses of a national park in Sri Lanka, respectively.

Discussion: The substitute method may adequately value quintessential use values by their substitutive counterfactual. But often products are less close substitutes than they appear at first sight. For instance, rural households in the Andes often use firewood as an inferior energy source, for example, for prolonged cooking, and more expensive kerosene or gas for light and quick heating needs, for example, boiling water (Wunder, 1996). Often local people have no economic means whatsoever to obtain the expensive industrial or urban substitutes that the valuation studies suggest. The alternatives that people de facto turn to in cases when natural resources run dry are often far from these 'luxury' suggestions. For instance, lacking local firewood supply may lead to the burning of dung, to steeply increasing firewood collection times from remoter sources and to an overall lowering of energy consumption – rather than a switch to gas, kerosene or soft coke. Another classical mistake is when locally non-marketed medicinal plants are being valued by the price of pharmaceutical substitutes, which are so expensive that poor rural households would never be able to buy it – thus applying an inadequate counterfactual for valuation. Care is thus necessary when using the commercial value of modern substitutes, because potentially gross over-valuations of natural values can occur, which lead to unrealistic results regarding the importance of extractive activities and about the economic values that forests and wildlands generate.

Embedded time and other inputs

Imagine firewood in a forest-near village is highly abundant, and thus fetches a zero 'resource rent' (in other words, an open-access raw product in its natural setting). No development scenario would realistically alter its supply (see Box 8.1). However, the value of already collected firewood is never zero: the labour time used for collection sets a minimum value for its 'kitchen-gate' value. For instance, Chopra (1993) in his valuation of Indian deciduous forests used embedded labour collection time in part to value firewood and other NTFPs. If processing the firewood on a larger scale, for example, for making charcoal, required a chainsaw, then beyond the operator's labour time, more capital costs and the cost of gasoline would also be embedded in the output value.

Households typically distribute their labour and other inputs in ways that equals marginal returns from different activities – or at least does not fall under a certain minimum. Locally paid wages (in other words, not the national minimum wage!) could serve as 'shadow values of labour', which we can use to price the firewood collection time. Other inputs, typically of raw materials or capital, can also be computed and added to the minimum price. In fact, household economic models can, based on information about physical production inputs used and returns from other activities, help computing implicit output prices. This production function approach to output pricing can at least serve to double-check the validity of other pricing methods – see the Campbell et al (2002) case study for Chivi district in Zimbabwe.

Discussion: Labour time and embedded input valuation is key to understanding rural livelihoods, but labour is also challenging to measure robustly: rural people tend to multi task (for example, collect firewood when returning from agricultural field), shadow costs differ across labour types (for example, between skilled and unskilled men, women and children) and seasons (for example, harvesting versus between-harvest seasons). In the PEN project, a strategic decision was made to not measure labour inputs, since it was thought that measurement efforts would be too resource-demanding. This generally precluded PEN researchers from using this valuation method. Finally, this conservative method with zero 'resource rents' assumes extreme abundance of natural products. As explained above, we are looking for the household's 'shadow price', which embeds both production costs and the utility of auto-consumption. What we implicitly assume with this method is that the shadow price is exclusively determined on the production side, so that the household demand side provides no value increment. In practice, this method thus often provides a lower value boundary, and is especially suited for inferior-type products.

Distant markets prices

Arguably the most common valuation error is to directly use urban-level market prices and multiply them by in-village production quantities to determine village-level subsistence values. This practice ignores that urban market prices include transport and marketing costs, whereas village-level value added may be only a fraction (see previous section). Moreover, distant market prices typically reflect purchasing power and levels of demand that are not present at local village levels. Even for a strictly speaking 'local' market, the valuation can become imprecise if it was to cover a variety of sub-sites with differences in resource availability and transport costs. Especially for bulky products such as firewood, charcoal, poles or fibres, errors in market location leads to glaring over-valuation errors.

Knowing the value chain, transport costs and margins of the product in other villages and close-by markets, one might possibly make corrections of the distant market price to arrive at a pseudo local price. These can also be estimated through surveys of value chain actors. In other words, if we find from 3–4 of these market studies that firewood tends to increase 0.10 shilling and charcoal 0.05 shilling per 50 km of transport distance to the market, then we can use these value increments in reverse: deducting market value added, what would have been the farm-gate value in our village?

Discussion: This method can make a useful complement only when product commercialization is a realistic counterfactual to direct use. It is also an applicable method of last resort when one has to validate uncorrected remote market information employed in an inadequate way; it is currently also being used for that purpose in the PEN project.

But distant market pricing would not be applicable when villages are extremely remote: transport costs would then come to exceed resource values, thus leading to negative imputed farm-gate prices. This would explain in economic terms why non-traded price bands are wide, and why product commercialization from our village of interest was not realistic in the first place.

Although we have described several methods individually, none may fully do the trick of delivering the exact desired value. Conversely, it is also seldom that a product can be simultaneously valued by all six methods; often by default only some methods are feasible. Combining these feasible methods for cross-checks and balances can thus lead to much more consolidated estimates, using both economic theory and common sense. The analyst should also not hesitate to make corrections and computations, as long as the assumptions are presented to the public in a transparent way. Valuing subsistence uses thus also requires viewing resources from comparative angles, and using economic common sense in making adequate choices (see also Box 8.2).

Box 8.2 *Can all non-marketed forest products be valued?*

Any comprehensive environmental income study in rural settings in developing countries is likely to come up with a long list of products used for a large number of purposes. Though it may be possible to accurately value many of these products, there may also be many pitfalls (for example, Adamowicz et al, 1998, Sheil and Wunder, 2002). Therefore, it may be necessary to try different techniques. As illustrated in this book, valuation may require careful planning, thorough data collection, continuous data quality control and opportunities to return to field sites to check suspicious data. And it may be necessary to pay particular attention to products that are of importance to most households' subsistence production, such as firewood, to ensure accurate value estimates for such products.

For some products, such as grasses and herbage consumed by grazing livestock, it may not be possible to directly obtain own-reported values. This is a challenge when one is working in an area where such products constitute an important component in household livelihoods, for example, in mixed agricultural systems reliant on grasslands for providing feeds to large livestock. For such products, one may be able to generate value estimates by combining own data with data from the literature.

There are also products that are impossible to value quantitatively. These may include sacred goods (Adamowicz et al, 1998), such as wooden religious artefacts, which people are unwilling to substitute, or in some settings the value of water consumption (Cavendish, 2002). Studies based on the PEN prototype questionnaire (PEN, 2007) generally do not record sacred goods or water consumption.

Checking data

The first three methods in the previous section rely directly on household's 'own-reported' values. The last three methods are estimates, in the sense that the outside analyst makes key assumptions regarding the applicability of imputed values. But there are many potential pitfalls, for example, when households feel obliged to provide answers to our contingent valuation questions even though they personally find it almost impossible. So, we need to check data quality: can we trust households' own-reported and our own analytically imputed values to provide valid and reliable measures? See also Chapter 12 for a general treatment of data checking.

A first step is to calculate basic distributional statistics (minimum, maximum, mode, median, mean, standard deviation) for unit values at product level (Cavendish, 2002). If households have provided us with valid data, we would generally expect:

- **Low dispersion in unit values**: For products with stable prices, we would expect standard deviation lower than the mean; and similar mean, mode and median values. The value band – in other words, the range within which values are estimated (determined by estimated minimum and maximum values) – should not be too wide, empirical evidence suggests that the value band for products with aggregated unit values with acceptable properties, is typically three to six times the standard deviation (Olsen, 2005; Rayamajhi and Olsen, 2008; Uberhuaga and Olsen, 2008). Products subject to fluctuating prices, for example, products with large seasonal price differences, such as pre- and post-harvest or heterogeneous products, will exhibit higher variation in unit values (see above for sources of price variation).
- **Homogenous standardized unit values**: Product values per SI unit (International System of Units, in other words, the modern metric system of measurement, including kilograms) should be similar across local units of measurement. Therefore it is useful to establish the relationship between local and SI units during fieldwork. Many conversion rates between local and SI units are product-specific (for example, how many kilograms of a particular fruit in a standard-sized basket?). Often many environmental products are used simultaneously, with various local measurement units. There may be much seasonal variation in product availability, and in some cases only few observations. Hence, conversion of local into SI units (which is needed if the physical quantities have to be compared across cases) is demanding, and should be explicitly planned for and continuously undertaken alongside value-data collection in the field.

- **Logical value ranking and correlations**: The value of processed products should be higher than for unprocessed materials (for example, charcoal will be more valuable than firewood), and similar products should have similar values (for example, different types of leaves used for the same purpose).

The few published studies using the above checking approach (Cavendish, 2002; Rayamajhi and Olsen, 2008; Uberhuaga and Olsen, 2008) showed unbiased own-reported values with satisfactory properties, which could hence serve for aggregation into product-level price estimates. Data checking can also include some aggregate common sense considerations. However, some valuation problems are commonly encountered:

- **Product size**: A particular product can exhibit variation for natural reasons, for example, species of mammals, fish or bamboo vary naturally in size. This can be dealt with through using more finely graded product categories – for example, registration of bamboo species – or by recording individual product details – for example, species and weight of hunted mammals.
- **Product quality**: Some product characteristics may not surface in interviews, for example, firewood may be wet or dry, or species composition may vary across loads. Again, this can be overcome by using more finely graded product categories.
- **Spatial and temporal variability**: In large study areas, own-reported values may vary due to differences in, for instance, transport costs and resource access. Some product values may also exhibit marked seasonality, for example, firewood may be valued higher in the winter, or fruits may be of low value in peak harvesting season. Keeping records at individual village level and collecting data across all main seasons should allow for analysis of such issues.
- **Few observations**: Many products may be encountered only once or a few times during a survey period. Data checking is hard for these products, but usually their share in total household income also remains low. In isolated hunter-gatherer communities where households collect a large number of products rarely, ignoring products with few observations may lead to underestimation of forest income. Additional information on such products can be collected through focus group discussions.

A systematic bias in the own-reported value data is problematic, as this would result in price data that do not reflect the true assigned value of products. The reason might be strategic responses on behalf of the surveyed population, such as wanting to appear poorer or underplaying the economic importance of illegally harvested products. Systematic bias should preferably be limited during fieldwork, for example, through clearly stating purposes of research and establishing good relationships to respondents (see Chapters 10 and 11). In general, checking data should already be done in the field, not at data entry stages. This allows the immediate cross-checking of suspicious estimates with households and informants.

Conclusions

In developed economies, non-market valuation is basically limited to the field of externalities and public goods. However, in developing countries, especially in rural areas, many products consumed at the local level do not enter the marketplace, or only do so partially. This is due to a number of structural obstacles and imperfections in output, factor, credit and insurance markets.

Valuing non-marketed products in rural tropical livelihoods is thus important in order to get a holistic view of household welfare and understanding the day-to-day decisions households make. Economic values reflect local scarcity and scenarios for the alternative use of resources at the margin of currently observed patterns. These do not necessarily integrate all the broader welfare considerations we as human beings are concerned with. But they provide good guidelines for how to optimize resource use at the margin of larger societal trends.

As economic analysts, we will necessarily want to 'compare apples with oranges', in other words, obtain a reductionist common monetary yardstick for ranking physical quantities that viewed in isolation would say little about household welfare outcomes. Failing to do so may misguide policy and project interventions, by ignoring the hidden harvest from multiple subsistence-oriented resource-extractive activities (Campbell and Luckert, 2002).

Determining these economic values, however, is not always easy; people's preferences, production functions and decision-making parameters cannot be read in an open book. Valuation normally requires some economic reasoning, an eclectic approach, a good portion of common sense and also some pragmatism. In the above sections, we described six different specific methods with their respective pros and cons, and how they could be creatively combined so as to cross-check value estimates from different angles and perspectives. Some of these rely on self-reported household values; others are analytical estimates. Under scenarios of imperfect information, setting upper and lower boundaries by triangulating different subsistence valuation methods may be the most promising approach. The relevance of methods may vary substantially across the subsistence products in question. There are often also strong spatial dimensions to natural resource values: implicit prices for one product may differ substantially between the forest, farm-gate and urban marketing levels.

Moreover, checking the value data's statistical and other properties can also provide important insights into the validity of findings. A careful product-level empirical analysis can reveal errors. But data checking should also include some aggregate common sense considerations. If we visibly perceive that villagers spend two thirds of their time on an activity we have valued at 10 per cent of all household income, we know that we have probably come to make an undervaluation somewhere. Or if half of the resulting household incomes in a village, for instance, proves to be concentrated in the subsistence use of wood fuels, we know that an error must have occurred, since the nature of human needs and composition of household spending elsewhere in the world would not justify such a concentration in household consumption. We have probably then chosen a wrong valuation method for firewood and should go back to have a second look. In other words, valuation is probably best conceived as an iterative procedure, where a process of trial and error will lead the analyst to reasonable estimates.

Key messages

- In rural areas of developing countries, especially remote settings with limited market access, the extent of non-marketed production can be substantial. Assigning inadequate values to these products can lead to major misunderstandings about local welfare and resource-use dynamics.
- We presented a prioritized list of six different methods to value subsistence goods, drawing on either household self-reported data or externally derived economic estimations. This list should be used eclectically, according to the specific case in question.
- For arriving at adequate value estimates, we recommend a thorough empirical check on household-reported value data, and the use of different subsistence methods and economic common sense to cross-check the results.

References

Adamowicz, W., Beckley, T., Macdonald, D. H., Just, L., Luckert, M., Murray, E. and Phillips, W. (1998) 'In search of forest resource values of indigenous peoples: Are non-market valuation techniques applicable?', *Society and Natural Resources*, vol 11, pp51–66

Adger, W. N., Brown, K., Cervigni, R. and Moran, D. (1995) 'Total economic value of forests in Mexico', *Ambio*, vol 24, no 5, pp286–296

Campbell, B. and Luckert, M. (eds) (2002) *Uncovering the Hidden Harvest: Valuation Methods for Woodland and Forest Resources*, Earthscan, London

Campbell, B. M., Jeffrey, S., Kozanayi, W., Luckert, M. K., Mutamba, M. and Zindi, C. (2002) *Household Livelihoods in Semi-Arid Regions: Options and Constraints*, CIFOR, Bogor

Cavendish, W. (2002) 'Quantitative methods for estimating the economic value of resource use to rural households', in Campbell, B. M. and Luckert, M. (eds) *Uncovering the Hidden Harvest: Valuation Methods for Woodland and Forest Resources*, Earthscan, London, pp17–65

Chopra, K. (1993) 'The value of non-timber forest products: An estimation for tropical deciduous forests in India', *Economic Botany*, vol 47, no 3, pp251–257

Costanza, R., d'Arge, R., de Groot, R., Farber, S., Grasso, M., Hannon, B., Limburg, K., Naeem, S., O'Neill, R., Paruelo, J., Raskin, R. G., Sutton, P. and van den Belt, M. (1997) 'The value of the world's ecosystem services and natural capital', *Nature*, vol 387, no 6630, pp253–260

Gunatilake, H. M., Senaratne, D. M. H. A. and Abeygunawardena, P. (1993) 'Role of non-timber forest products in the economy of peripheral communities of Knuckles National Wilderness Area of Sri Lanka: A farming systems approach', *Economic Botany*, vol 47, no 3, pp275–281

Hanley, N. and Barbier, E. (2009) *Pricing Nature: Cost-Benefit Analysis and Environmental Policy*, Edward Elgar, Cheltenham

IIED (International Institute for Environment and Development) (2003) *Valuing Forests: A Review of Methods and Applications in Developing Countries*, International Institute for Environment and Development, Environmental Economics Programme, London

Mekonnen, A. (2000) 'Valuation of community forestry in Ethiopia: A contingent valuation study of rural households', *Environment and Development Economics*, vol 5, no 3, pp289–308

Mitchell, R. C. and Carson, R. T. (1989) *Using Surveys to Value Public Goods: The Contingent Valuation Method*, Resources for the Future, Washington, DC

Olsen, C. S. (2005) 'Trade and conservation of Himalayan medicinal plants: *Nardostachys grandiflora* DC and *Neopicrorhiza scrophulariiflora* (Pennell) Hong', *Biological Conservation*, vol 125, no 4, pp505–514

PEN (Poverty Environment Network) (2007) 'PEN prototype questionnaire, version 4', CIFOR, Bogor, www.cifor.cgiar.org/pen, accessed 2 November 2009

Rao, V. (2001) 'Celebrations as social investments: Festival expenditures, unit price variation and social status in rural India', *The Journal of Development Studies*, vol 38, pp71–97

Rayamajhi, S. and Olsen, C. S. (2008) 'Estimating forest product values in Central Himalaya: Methodological experiences', *Scandinavian Forest Economics*, vol 42, pp468–488

Sadoulet, E. and de Janvry, A. (1995) *Quantitative Development Policy*, Johns Hopkins University Press, Baltimore, MD and London

Sheil, D. and Wunder, S. (2002) 'The value of tropical forests to local communities: Complications, caveats, and cautions', *Conservation Ecology*, vol 6, no 2, pp9–16

Uberhuaga, P. and Olsen, C. S. (2008) 'Can we trust the data? Methodological experiences with forest product valuation in lowland Bolivia', *Scandinavian Forest Economics*, vol 42, pp508–524

Whittington, D. (1998) 'Administering contingent valuation surveys in developing countries', *World Development*, vol 26, no 1, pp21–30

Wunder, S. (1996) 'Deforestation and the uses of wood in the Ecuadorian Andes', *Mountain Research and Development*, vol 16, no 4, pp376–381

Chapter 9

Preparing for the Field: Managing AND Enjoying Fieldwork

Pamela Jagger, Amy Duchelle, Sugato Dutt and Miriam Wyman

Experience is one thing you can't get for nothing.
Oscar Wilde (1854–1900)

Introduction

Embarking on fieldwork is for some the most exciting and challenging part of the research process. How fieldwork is organized, and how researchers and their teams present and conduct themselves, can have a significant impact on data quality and research team members' well-being, happiness and health. Before embarking on fieldwork, considerable preparations should be in place: collection of good background information (Chapter 5); the sampling strategy (Chapter 4); hiring a research team (Chapter 10); and designing and pre-testing questionnaires should be completed (Chapters 6, 7 and 10). Now the time has come to start collecting data. The purpose of this chapter is to discuss practical issues that will help researchers cope with and enjoy fieldwork. This includes suggestions for strategies that can help with navigating challenging political and cultural situations and practical advice on doing fieldwork. Figuring out where to live, what to eat and drink, and how to stay healthy and safe are critical aspects of a productive and positive field experience.

Context matters! The importance of political and cultural context

Rushing to get to the field is a mistake many researchers make. Before setting foot in the village, there is a considerable amount of research and administrative legwork to be done. In addition to finalizing research instruments, hiring and training a research team, pre testing data collection instruments and selecting study sites, researchers need to learn about the political and cultural context of the study area. First impressions can make a very big difference to success in the field, and preparing carefully before starting fieldwork will have a positive influence on the integrity and quality of the data collected.

Political context

We emphasize three political issues that need attention: (a) understanding formal and informal hierarchies and approval processes; (b) acknowledging the special case of natural resources and issues linked to resource access and use; and (c) knowing the political and economic history of the area (Magolda, 2000; Gubrium and Holstein, 2001; Ergun and Erdemir, 2010).

Understanding formal and informal hierarchies and approval processes

In most countries there are both informal and formal hierarchies and procedures that researchers are advised to observe. Most countries have a research approval process that requires researchers to obtain research permits from a national-level entity prior to undertaking research of any kind. In addition, there might be procedures for obtaining permission to go into the field (for example, getting written permission from the district-level police or military authority in the area). These processes can be very bureaucratic and complex and take several months; leave sufficient time for obtaining approvals before heading to the field. It can be tremendously helpful to have a local collaborator who is familiar with the approval process and who will vouch for your credibility. In many countries, having a local sponsor for your research is required.

Once formal research approval at the national level has been obtained, directly inform people who should be made aware of the research project and fieldwork plans including: provincial or district officials; military outposts; non-governmental organizations (NGOs) undertaking activities related to your research; collaborators and colleagues at academic or national research institutions; natural resource management authorities, and so on. This is beneficial for two reasons. First, important resources or key informants may be uncovered (for example, someone who has digitized village boundaries or a key informant who has moderated disputes over forest resources). Second, making sure that people are aware of the research team's presence and movements in the field should contribute to its relative success and safety. There may be several levels of informal and formal hierarchies to work through before going to the villages where you will conduct your study. For example, in countries with decentralized governance systems, it may be important to make research objectives known to multiple levels of government officials. Having a letter of introduction that briefly describes the research team leader, the research project and objectives, the research team and the specific areas in which research will take place is a good idea. Letters written by respected in-country collaborators or institutional partners are particularly helpful. Put the letter on official letterhead, leave a space in the salutation line so it can be personalized and make sure to print sufficient copies.

Acknowledging the special case of natural resources and issues linked to resource access and use

Natural resource management is fraught with political complexities including: land redistribution; contested land and forest tenure; unclear or overlapping property rights systems; and conflicts between local resource users and outsiders and/or forestry and other officials. For example, in Bangladesh, land tenure is contested, making questions regarding property rights potentially difficult to broach with communities and households (Box 9.1). It is critical to have knowledge of the study area context, in particular knowledge of conflicts. Erroneous assumptions about access, use, management, distribution, and so on, of natural resources and products can compromise the research team's credibility in the field. Grey literature produced by government agencies, donors and NGOs is often a valuable source of information. Interviewing a diversity of key informants prior to arriving in the field is an important source of knowledge – try to identify a sample of key informants that will provide a diverse set of views, and who are likely to identify issues that might not be highlighted in government, donor or NGO reports. Talking to researchers that have previously worked in the area can also be a useful source of information.

Collecting data regarding illegal activities and navigating relationships between forestry or environment officials and communities are significant challenges associated with collecting valid and reliable data on forest and environmental incomes. Ensuring confidentiality is a critical aspect of successfully collecting complete information. Many forest and environmental goods are harvested illegally, making respondents nervous about revealing if, what and how much were harvested. While in the field, emphasize the aggregation of data to respondents and officials at all levels (from village leaders to district chairmen), respondents should be confident that reported data will never be used to draw attention to particular activities undertaken in their household. A basic understanding of the politics and economics of high value resource extraction beyond the village boundaries is also useful: local officials, high-level politicians, military officials, and so on, might be involved in both legal and illegal extractive activities.

Box 9.1 *Politics surrounding land tenure: Forest officials and their relationship with communities*

Ajijur Rahman

Understanding local dynamics of land tenure and community relations with forest officials are examples of understanding the political context of your field site. Most land in the uplands of eastern Bangladesh is owned by the state – although people use state land for their subsistence needs, they do not have any permanent or long-term rights to the land. This lack of tenure security promotes the practice of shifting cultivation, which is the main driver of deforestation in the area. Weak tenure security also limits access to formal credit, as small farmers cannot supply required collateral, forcing poor farmers to get loans from local moneylenders at high interest rates. This means less investment in good land management practices. This tense situation requires that researchers need to establish good relations with local communities before asking about sensitive tenure issues.

Understanding relations between small farmers and official authorities is also important. In Bangladesh, the Chittagong Hill Tracts Forest Transit Rules (1973) and subsequent administrative orders regulate the harvesting and marketing of timber and other forest products available from private growers. The rules require people to get written permission from government offices before harvesting and transporting forest products, especially marketed timber. As such permits are not issued to small farmers, tree growers are compelled to sell timber to local traders at low prices, discouraging private tree growing. In addition, forestry officials are located in sub-district headquarters, far from the farmer, and have limited understanding of rural livelihoods and constraints. Understanding these relationships is important for the researcher working in this study area: the local forest officer, often among the first people consulted to learn about forest management in the region, is not likely to provide an accurate picture of obstacles to obtaining forest income. Further, if researchers are perceived to be too closely allied with forest officials, if introduced in villages by these officials, local trust may be compromised.

Local perceptions of research team interactions with officials and organizations influence the quality of collected data. Asking local natural resource management officials to introduce the research to village leaders and community members means that the research may automatically be associated with the introducing organization or personnel. Any resentment or hostility against the organization or person could then be directed towards the research.

Households may not want to share information for fear that it will pass to the organization (for example, regarding illegal harvesting activities). If someone to make introductions to village leaders is needed, try to find a government official, local leader or NGO representative who is neutral with respect to natural resource allocation and use.

Knowing the political and economic history of the area

Make sure the research team is familiar with the political and economic history of the study site; political, social and economic relations are often shaped by the history of an area. There is a wealth of information to be picked up in books and articles about almost all countries and regions – search libraries and the internet. Often local printing presses publish books that have a limited distribution outside of the country. Bring history books to the field and ask respondents about major events and find out how the oral histories of elderly people in the village compare with academic accounts of events.

Cultural context

Most researchers are not from the village or area where fieldwork is conducted and are thus perceived as outsiders. Familiarization with local customs and language, and establishing trust early on will help in overcoming potential barriers of entry and encourage greater willingness to share information among respondents.

Understand the culture

Cultural differences are typically related to ethnicity, nationality, age, race, gender, religion, caste and socio-economic status. Respondents may view cultural differences as a threat and be reluctant to give information if they: feel vulnerable to legal action; feel intimidated by the researcher; feel other community members could use the information to further institutional agendas or legitimize social inequalities; have insecurities regarding interviewing across class, gender, race or ethnic lines (Adler and Adler, 2001; Briggs, 2001; Ryen, 2001; Shah, 2004). Researchers need to figure out how to overcome, or at least cope with as many of these potential barriers as possible.

We propose the following list of best practices for demonstrating cultural awareness and overcoming cultural barriers:

- Understand local fears, anxieties and sources of pride.
- Be humble and do not gratuitously display wealth.
- Dress respectfully, a good rule is to be dressed slightly more formally than respondents.

- Address fears and concerns with empathy.
- Work hard to communicate with respondents even if you do not speak their language.
- Train all team members to be culturally aware and avoid stereotypes.
- Match enumerators and respondents by cultural compatibility (for example, language skills and gender issues).
- Use cross-cultural teams, including a gender-balanced enumerator team.

Enumerator training (see Chapter 10) should include sessions on conduct in the field and coping strategies for dealing with uncomfortable situations, such as asking about payment for time spent responding to questionnaires, dealing with conflicts between households in the village and even how to deal with threats of violence. Even when enumerators speak the same language as respondents, there are barriers to overcome.

Establish trust

Data quality and the research team's overall field experience are strongly influenced by the level of trust established with the study villages. Building trust and rapport with respondents means integrating into community life: attending community festivals and sporting events, walking around the community and spending time getting to know families, helping with community projects and maybe even offering to give lessons (for example, English tutoring for students). Establishing trust with all groups is essential, including marginalized people. For example, in some cultures, women are excluded from formal meetings; this should be addressed when requesting village leaders to bring a representative group together for focus group discussions.

Learn the local language

Being able to speak directly with respondents is a tremendous asset. Invest a few weeks or months in intensive language training and learn the basics of greetings and showing gratitude. Knowing a few local proverbs can help break the ice with respondents. Learning enough vocabulary to follow an interview (in other words, agriculture and forestry terminology) is extremely helpful for cueing enumerators to probe further when interesting or unusual responses are given. It is essential to work with enumerators that have a strong command of the local language. Be aware that dialects differ widely. If you are working across a relatively large geographic area with diverse linguistic groups, you should consider having more than one research team. Alternatively, you can hire translators to work with enumerators. Conducting surveys using a translator is a sub-optimal situation. Translation lengthens the time of the interview, reduces the validity and reliability of data (important information is lost in translation), and puts respondents and enumerators on edge as they both have to work harder to understand the meaning of responses. Body language also tells a lot, such as facial gestures, rapport and demeanour. With a translator, having someone who not only communicates in the local language, but who also understand local conditions, customs, practices, and so on, can make things easier. Finally, be cautious about importing one set of linguistic and cultural assumptions into another when interviewing between cultures. Even within the same culture, meanings that seem clear to the interviewer may not be clear to the respondent.

Be transparent

Make the research process transparent to respondents. They will be naturally curious to know why their village or household was selected for the study. Meeting with community leaders and holding a community meeting that anyone can attend is a good way to inform people of the objectives of the research and how and why they were selected. A random sample of households can also be drawn at such a meeting, demonstrating that household participation is by chance rather than through connections. Respondents may also have questions regarding the confidentiality of the information they provide, how the information will be used and how they will benefit from participating. Be open and clear from the start about the purpose of the research. To ensure transparency:

- Have the lead researcher visit all households to offer an explanation of the research; this will underscore the importance of the survey work and may encourage greater participation and higher quality data collection.
- Offer respondents an opportunity to express concerns or ask questions about the content of questionnaires, the interview process, or how data will be used.
- Conduct interviews in a place that the respondent is comfortable in.
- Acknowledge that the respondent does not have to answer questions that make them uncomfortable.
- Think about the timing and ordering of personal questions.

Several factors influence willingness to participate in a study, including: timing (for example, if it is planting or harvesting season people are busy and may not want to commit time to respond to questionnaires); the level of research fatigue in the study site; whether or not a gift or compensation for participation is offered; and the general level of interest in the research team and purposes. Be respectful of people's time and commitments, through careful planning it is possible to anticipate times or days (for example, holidays) that are not ideal for administering questionnaires.

There is an ongoing debate regarding the practice of giving gifts or money as an incentive to participate in survey research (Lynn, 2001; Wertheimer and Miller, 2008). The general rationale for giving gifts is that people are busy and should be adequately compensated for their time or contribution to a study. Most controversial are cash payments directly to households. The criticism is that directly paying someone to respond to your survey is a type of coercion; scientific integrity may be compromised by commodifying a practice that should be based on altruism. Poverty Environment Network (PEN) researchers were encouraged to give practical gifts (for example, salt, sugar, matches, soap or pencils) to participating households. The general advice was to give a gift valued at roughly the daily wage rate for unskilled labour.

Restitution (reporting back to respondents)

At the end of fieldwork, it is a sign of professionalism and respect to share preliminary findings with local communities and partner organizations that helped support the research. Too often field researchers extract information from local communities and leave suddenly with little to no closure or follow-up. In this section, we discuss techniques for giving proper closure to the research process through disseminating preliminary results and thanking local communities and partner organizations before going home. This process is sometimes referred to as 'restitution'.

Although there are various phases in the research process in which researchers can engage local stakeholders, restitution may be the phase where it is easiest and most effective for researchers to engage local people (Kainer et al, 2009). Returning preliminary results to communities and partner organizations during or at the end of field research serves the dual purpose of sharing information gleaned during the research process, while allowing researchers to validate preliminary findings based on feedback from local stakeholders. Dissemination of research results can take a variety of forms, including short presentations, interactive workshops, brochures, maps, radio and photo albums, with researchers limited only by their creativity, available resources and knowledge of audience-appropriate methods for sharing scientific information (Shanley and Laird, 2002; Duchelle et al, 2009; Kainer et al, 2009).

The extent to which field researchers will be able to disseminate preliminary results will depend on the timing of their research. For instance, the advantage of research projects that require multiple visits to communities is that they allow researchers to develop relationships with the communities where they work; each field visit can be used as an opportunity to share select preliminary findings at a community meeting. Such information sharing throughout the research process clearly shows the researcher's appreciation for community involvement and treats community members as research partners and not simply respondents.

Information sharing is generally welcomed by local stakeholders, allows for mutual learning and provides an important vehicle for researchers to thank local communities and partner organizations as fieldwork comes to a close.

In addition to dissemination of research results, there are a variety of other ways to thank local communities and partner organizations for their supporting role in field research. These activities are again determined by the researcher's creativity and resources, and by what is appropriate in the local context. For example, at village meetings during the final field visit, researchers can present personalized certificates to participating households to publically recognize and thank them. Other ideas include hosting lunches or parties. Such gestures clearly show researchers' appreciation for the time and energy that local people give to field research, keeping the door open for local stakeholders to want to engage and collaborate with researchers in the future.

The practicalities of life in the field

Researchers should spend as much time in the field as possible: it can be enjoyable, it can generate a lot of contextually relevant information and, perhaps most important from the perspective of the topic of this book, the data quality will be much better (more accurate and complete). Getting high quality data requires detailed checking and quality control, and should not be left to enumerators or research assistants. This section focuses on practical aspects of fieldwork organization, including the implications for researchers' health and safety.

Where to live?

Beyond avoiding areas with armed conflict or drug wars, where to live was probably not a primary concern during study site selection. If fieldwork is conducted in one village, or several villages within fairly close proximity to one another, living in the village is potentially a good option: researchers are integrated into the community and likely to build a high degree of trust with respondents. This can facilitate the data collection process tremendously, providing opportunities to see things from an anthropological perspective, to groundtruth trends observed in the data and to collect information about activities that might not be easily observed otherwise. For example, in many forested areas, illegally harvested timber is picked up by traders and transporters in the middle of the night. By being around all the time, it is possible to learn about social, political and economic processes that might not be obvious – while minimizing research costs at the same time.

There are also drawbacks to living in a village: it is easier to get involved in village politics and more difficult to get privacy. Researchers boarding with a local household, staying in a local motel, living at an NGO or college guesthouse or building their own dwelling close to the village leaders may find that respondents develop perceptions about possible allegiances. As we have discussed, perceived political or ethnic allegiances can be detrimental to data collection. If respondents feel that one group is favoured over another, then trust is compromised. The challenge for the researcher living in the field is to navigate these relationships in a context that is culturally different and operates with a different set of social norms than the researcher is used to. While status as an outsider allows for a degree of tolerance related to social and political 'errors', researchers should strive to gain awareness of local customs relatively quickly. Limitations on privacy affect people differently. Researchers should expect to be the focus of a lot of attention, particularly as they arrive and get settled. Children in particular will be curious. Over time, however, village members are likely to decrease their level of interest.

Many of the practical details of village living are beyond the scope of this chapter. To help guide decisions on where to live, the following issues should be considered:

- Will you build your own house, camp or board with a local family?
- Where will you obtain food and who will do the cooking?
- What is the source of fresh water and what are the bathroom/shower facilities you will use?
- Where will you store your questionnaires and research notes?
- Where will you store your valuables (money, mobile phone, computer, and so on)?
- What is the reliability and cost of transportation in and out of the village?
- Do you have access to health care facilities?
- Are you prepared to deal with vermin, snakes, insects, and so on, that might be an issue in your field site?

Food in the field

It is common for people to welcome researchers to their village or home by offering food and drink. These may be exotic, such as dried white ants in groundnut sauce, served with matoke (cooking banana); 'bamboo chicken', also known as iguana; tiny fried frogs; or python preserved in locally made brew. Undoubtedly fieldwork pushes researchers to the limits of their epicurean comfort zone – either from eating the same meal of pounded cassava for days on end or from being offered very exotic fare. Remember that food is strongly tied to culture: offering food and drink is a form of hospitality across many cultures and food may be the only thing that the most humble households have to offer. To refuse to eat offered food in a respondent's household may be perceived as extremely rude, depending on the cultural context and may have implications for how researchers are perceived in the household or the larger community. Researchers collecting data on environmental income should be particularly willing to try wild foods. Part of understanding local culture, tastes and preferences is getting to know a culture through food. So unless religion, culture or health forbids, eat it!

Days in the field can be incredibly long. If working in a community where food or drink is not available, finding a way to eat enough to keep going through the day can be a major challenge. In some cultures it is rude to eat in front of people that are not themselves eating. A useful strategy is to support the local economy by purchasing food, such as fruits, roasted maize or cassava, sodas, ground nuts, roasted meat or a wide variety of other foods that are often available in households or village trading centres. Purchase from a diverse set of suppliers and do not be too tough on the bargaining – this will support the local economy and serve to maintain good relations with hosts.

Partaking in local social activities is a great way to break the ice with respondents. For example, in sub-Saharan Africa, many cultures have one or two forms of locally made alcohol and drinking is often a social activity. While drinking is a great way to get integrated into a community, be aware of the risks of drinking alcohol produced under village conditions. Remember to be respectful of village settings where alcohol is not consumed.

Staying healthy

Staying healthy in the field is a challenge whether living in a village or commuting daily from a nearby town. Research teams work long days, frequently under harsh physical conditions. There is great potential to get physically run down, increasing susceptibility to illness and disease. The most likely health problems in the field are: dehydration; sunstroke; water-borne illnesses including dysentery; and mosquito-related illnesses. Lower probability health risks include snake bites, stings, leeches and accidents. It is good to have information on localized epidemics and to take the necessary precautions. Make appropriate preparations before fieldwork, get the recommended vaccinations and prophylactics, and have appropriate health insurance for all team members. Private insurance is available in many countries. Basic first aid training for all team members is a great idea and at least one member of the team should have first aid supplies in the event of an emergency. Driving is probably the activity that entails the highest risk in the field. When renting a vehicle (and driver), set clear rules, such as maximum speed limits.

Pay particular attention to the health of research team members. Just because an enumerator has had malaria several times before, it does not mean it is easier for her to deal with. Encourage team members to take precautions including accepting food and drink that have been prepared with some attention to food safety, sleeping under a mosquito net, avoiding sunstroke, and so on. When enumerators fall ill, adjust the research programme accordingly. Overworking enumerators who are having health problems leads to low morale and ultimately compromises the data collection efforts.

Tips for maintaining good health in the field include:

- Be aware of local epidemics and take the necessary precautions.
- Make sure that a reliable source of clean drinking water is available or boil water and store it in manageable quantities; carry water into the field.
- Maintain personal hygiene, including washing hands on a regular basis.
- Store food in a safe and hygienic location.
- Always carry a first aid kit.
- Always carry identification, insurance and medical aid details.
- Make contacts with local doctors or hospitals to check out the assistance available in case of emergency.
- If someone gets sick, another person should take charge. Do not leave it to the sick person to decide, often they are unable to make rational decisions.
- Make contingency plans in the event that team members get sick.

Researchers are frequently called upon to assist with health crises in the villages they are working in. The two most common forms of assistance are providing transportation to the nearest health centre or hospital and providing financial assistance to households experiencing health crises. The most important advice we have is to treat village members equally. If transportation and funds to support the medical needs of one or two families are provided, be prepared to do so for others in need. While requests for transportation and financial assistance can be quite taxing and interfere with fieldwork schedules, researchers should be aware of the impact that helping out might have on people's well-being, and the goodwill the community will extend to you.

Personal safety

Personal safety is a major issue for all researchers, whether a foreigner or a national of the country in which work is taking place, whether a man or a woman. As a relatively well-off outsider, there is a good chance that respondents in the villages will perceive researchers as wealthy. Researchers should therefore at all times take appropriate precautions, for example, living with a trusted family. While there may be other issues tied to this (perceived allegiances and limited personal space as noted above), personal safety and the safety of belongings will be higher than when living alone. This will also benefit families in the communities through rent and will help the researcher to assimilate into the community. Take into account that 'Murphy's Law was written in the tropics'; many plans can and will go wrong, so make some contingency arrangements whenever possible, for example, the research team should agree on basic safety routines: what do we do if the car breaks down, a research team member is attacked or gets sick, and so on.

To avoid awkward and potentially dangerous situations:

- Make living arrangements in a reasonably well-inhabited place with secure doorways and a sturdy lockable place to store all valuables.
- Surround yourself with trustworthy people; research assistants should ideally live in the same place, or at least within reasonable distance, for quick and easy communication if an unfortunate situation arises.
- Distinguish between genuine well-wishers and schemers or eavesdroppers who can pass off information on the location of valuable assets, or times when research team members might be alone and vulnerable.
- Be certain to understand the lay of the land in both the literal and figurative senses:
  - Know your way around – understand return routes and pathways when negotiating new locations.
  - Be aware of ethnic and political conflicts that the research may be at the heart of – for example, sympathizing with forest officials may put the research team under suspicion of households that engage in illegal logging.
- Make careful choices regarding study areas, avoiding/abandoning if possible those that are conflict-ridden or where conflict could emerge.
- Give full details of medical aid, next of kin, contact numbers, ID number, and so on, to the local partner institution or someone that you are working with (but who does not accompany you to the field). They should also get the detailed field trip plans (where and when).
- Keep the local headman, village chief, police station or other relevant authorities informed about the research team's stay and movements.
- Fieldwork should not be undertaken alone. There must be two people present, preferably three, of whom at least one should be male.
- Bring a mobile phone, if the area is connected.
- Take seriously any threats from individuals received at your study sites.

- Beware of participating in 'dubious' social gatherings (for example, those with lots of alcohol involved) after dark when alone and unprotected.
- Be careful when hitchhiking or offering lifts to strangers, especially when alone, in remote areas and/or after dark. The rule of thumb is to never give lifts to anyone not associated with the research.

Female researchers should exercise extra caution in the field (Box 9.2). Though you might wonder at times how you could ever find yourself alone (you are

Box 9.2 *In memory of Vanessa Annabel Schäffer Sequeira (1970–2006)*

Vanessa Sequeira, a Portuguese doctoral student, was conducting PEN research in Acre, Brazil, in Western Amazonia when she was brutally murdered in the field. Vanessa was alone in a remote part of her study area looking for a family to interview when she was attacked by a man who had recently been released from prison for committing a similar violent crime. While the crime against Vanessa was determined to be not in any way directly related to the research that she was undertaking, the shocking and devastating act was a terrible reminder of the fragility of life and the vulnerability of researchers in the field.

Vanessa had extensive experience in Amazonia and a bright future ahead of her. Before beginning her doctoral work at the University of Bangor, UK, and CATIE (Centro de Altos Estudios de Conservacion y Investigacion de Agricultura Tropical de Costa Rica), she worked for four years directing field research for the Proyecto Conservando Castañales in Madre de Dios, Peru, where she implemented a project for sustainable management of the Brazil nut, a regionally important non-timber forest product. In Acre, her doctoral research focused on the differences in forest dependence between colonist settlers and forest extractivists. In her preliminary analyses, she reported remarkable differences between the forest extractivist and colonist communities and was delightedly proving her original hypothesis wrong – it seemed that small producers who actually used the forest were better-off than those who had cut their forest down to raise cattle.

Vanessa had a gift for engaging local communities and she strongly believed in conducting research that could make a difference for tropical conservation and for the people with whom she worked. Vanessa was tremendously loved and respected by the Peruvian Brazil nut collectors in Madre de Dios, and in Acre she considered many of the extractivist families in the Agro-Extractive Settlement Project Riozinho-Granada close friends. Her friends and colleagues were inspired by her persistence, humility and great sense of humour. Vanessa was a light in the world of tropical conservation and development, and she is missed terribly (www.vanessa.sequeiras.net).

always at someone's house or being followed around by children), you might find yourself in an uncomfortable or potentially dangerous situation. The bottom line: live and work in the field in an open and trusting way, but exercise a reasonable level of caution at all times. In particular, think carefully about situations where you will find yourself alone and vulnerable.

Conclusions

Our main message is that fieldwork can be a wonderful enriching experience if researchers invest time in understanding the context of the fieldwork site and situation, and think carefully about how the research team can function effectively in the field. Our assumption, validated by the field experience of several PEN researchers, is that well-organized fieldwork is correlated with data quality. Beyond the quantitative data collected, there is a significant qualitative story that needs to be understood. Being comfortable in the field and having good rapport with respondents gives the best shot at understanding the complexities and nuances that underlie the quantitative data.

Key messages

- Do not rush into the field, first get to know the political and cultural context of the study sites.
- Think hard about where to live and how to organize the research so that all research team members are healthy and safe.
- Fieldwork is a life-changing experience – embrace it and have a great time!

References

Adler, P. A. and Adler, P. (2001) 'The reluctant respondent', in Gubrium, J. F. and Holstein, J. A. (eds) *Handbook of Interview Research: Context and Method*, Sage, Thousand Oaks, CA, pp515–535

Briggs, C. L. (2001) 'Interviewing, power/knowledge, and social inequality', in Gubrium, J. F. and Holstein, J. A. (eds) *Handbook of Interview Research: Context and Method*, Sage, Thousand Oaks, CA, pp911–922

Duchelle, A. E., Biedenweg, K., Lucas, C., Virapongse, A., Radachowsky, J., Wojcik, D. J., Londres, M., Bartels, W., Alvira, D. and Kainer, K. A. (2009) 'Graduate students and knowledge exchange with local stakeholders: Possibilities and preparation', *Biotropica*, vol 41, no 5, pp578–585

Ergun, A. and Erdemir, A. (2010) 'Negotiating insider and outsider identities in the field: "Insider" in a foreign land; "Outsider" in one's own land', *Field Methods*, vol 22, no 1, pp16–38

Gubrium, J. F. and Holstein, J. A. (eds) (2001) *Handbook of Interview Research: Context and Method*, Sage, Thousand Oaks, CA

Kainer, K. A., DiGiano, M. L., Duchelle, A. E., Wadt, L. H. O., Bruna, E. and Dain, J. L. (2009) 'Partnering for greater success: Local stakeholders and research in tropical biology and conservation', *Biotropica*, vol 41, no 5, pp555–562

Lynn, P. (2001) 'The impact of incentives on response rates to personal interview surveys: Role and perceptions of interviewers', *International Journal of Public Opinion Research*, vol 13, no 3, pp326–336

Magolda, P. M. (2000) 'Accessing, waiting, plunging in, wondering, and writing: Retrospective sense-making of fieldwork', *Field Methods*, vol 12, no 3, pp209–234

Ryen, A. (2001) 'Cross-cultural interviewing', in Gubrium, J. F. and Holstein, J. A. (eds) *Handbook of Interview Research: Context and Method*, Sage, Thousand Oaks, CA, pp335–354

Shah, S. (2004) 'The researcher/interviewer in intercultural context: A social intruder!', *British Educational Research Journal*, vol 30, no 4, pp549–575

Shanley, P. and Laird, S. A. (2002) '"Giving back": Making research results relevant to local groups and conservation', in Laird, S. A. (ed) *Biodiversity and Traditional Knowledge*, Earthscan, London, pp102–124

Wertheimer, A. and Miller, F. G. (2008) 'Payment for research participation: A coercive offer?', *Journal of Medical Ethics*, vol 34, no 5, pp389–392

Chapter 10

Hiring, Training and Managing a Field Team

Pamela Jagger, Amy Duchelle, Helle Overgaard Larsen and Øystein Juul Nielsen

---

You are only as good as the people you hire.
Raymond Albert Kroc (1902–1984)

Introduction

Even with a theoretically and empirically sound questionnaire, there are practical aspects of survey implementation that can significantly affect data quality and a research project's cost-effectiveness. Most household level socio-economic surveys involve the interaction of researchers, respondents and enumerators. Each of these interactions is an opportunity to collect high quality data but also presents opportunities for data quality to be compromised. Field researchers must pay explicit and serious attention to human resource development and management, project management and social capital building. This chapter addresses three aspects of field research that seldom receive focused attention: (a) hiring and training the field team; (b) managing and motivating the field team; and (c) putting into place good questionnaire management procedures. These skills help field researchers develop strategies to mitigate common problems including: inadequately trained enumerators; enumerator attrition; communication problems or conflicts between researchers, enumerators and respondents; missing or incorrect data filled in on questionnaires; and loss or theft of questionnaires. The limited available information on these issues is scattered across grey literature (for example, FAO, 1995; Hoare, 1999; PEN, 2007), websites (for example, NASDA, 2010; WCSRM, 2010), older books (for example, Devereux and Hoddinott, 1992; Poate and Daplyn, 1993; Puetz, 1993; Vemuri, 1997) or more rarely in peer-reviewed papers (for example, Ross,

1984; Whittington, 2002). See Chapter 11 for a discussion of operational criteria for assessing data quality, including biases that may arise when using enumerators.

Hiring a field team

The selection and training of enumerators is central to the quality of data collected. Unless a researcher has a relatively high level of proficiency in the local language and a relatively small sample size, she is going to need people to help implement field surveys. Putting together a good team, then training and managing that team are big challenges, particularly for first-time researchers. There are several critical factors that strongly influence the selection of enumerators including: level of education; desired language skills; prior experience with socio-economic surveys; local knowledge of the region or communities where fieldwork will be conducted; and research budget.

We distinguish two types of enumerators: *external* enumerators, who are university students or recent university graduates (local or non-local) or local professionals in non-governmental or governmental organizations, and *local* enumerators, who are often members of the community where the research is being carried out. The decision to hire external or local enumerators is not a trivial one. The main advantage of using external enumerators is that their education and experience is expected to improve the quality of data collected. External enumerators may already have relevant research experience and a high degree of familiarity with collecting complex socio-economic survey data. They are likely to require less training than local enumerators who have usually not collected field data before. Also, external enumerators may have extensive field and/or research experience in the study region, along with personal connections in rural areas, allowing them to help contribute to both sound data collection and the logistics of fieldwork.

The main disadvantage of hiring external enumerators is the cost: generally, external enumerators command higher wage rates and there are potentially high costs associated with transporting the field team to the study site. Logistical considerations should not be underestimated: time required to travel to remote survey villages can reduce the time spent in the study area collecting high quality data. One way to reduce time and transportation costs is to have enumerators spend weeks at a time in the study villages or to work there on a permanent basis throughout the survey. A likely second constraint associated with hiring external enumerators is a lack of knowledge regarding the study site, so extra time and energy will be needed to gain local trust and insight into local conditions.

The major benefit of hiring local enumerators is their in-depth knowledge of the local area. They can provide relatively quick access to local information not readily available to outsiders, including information regarding illegal activities, and they are generally very conversant about local conditions and customs. The use of local enumerators reduces costs of salaries and the logistics of arranging transportation and accommodation. This is particularly helpful when research sites are remote and/or households are scattered over large geographical areas. The disadvantage is that their educational skills are generally lower and therefore more time needs to be spent on training. Also, a potential limitation of employing local enumerators is that some information may not easily be shared among neighbours, such as the value of hidden assets or issues related to conflict over natural resources.

In hiring either external or local enumerators, it is critical to decide whether to rely on recommendations from colleagues (for example, social networks, professional colleagues, schools, churches or non-governmental organizations (NGOs)), or to undertake a meritocratic hiring process that includes advertising the job and interviewing the most qualified candidates. There are strong arguments for either approach. Asking local collaborators for enumerator recommendations can provide quick identification of strong candidates. However, the downside, especially in the case of local enumerators, is the risk that recommendations are motivated by family relationships, ethnic ties, and so on. Employing local enumerators introduces attractive opportunities for local people to generate significant income, which may induce locally prominent people to influence the selection process. The challenges of applying a merit-based approach to hiring enumerators are twofold: the process may be unfamiliar or inappropriate in some settings; and it may be difficult to evaluate local references. Regardless of the identification procedure, all candidates should be screened through interviews before initiating training.

Researchers with a large number of enumerators (three or more) may want to consider hiring a more experienced team member to help supervise the team. An enumerator supervisor should have more significant responsibilities and should receive a higher salary. Having someone that understands the local context well, can serve as an ombudsperson between researcher and enumerators, and can help check questionnaires in the field can be tremendously valuable. A good assistant can also be a valuable partner in discussions regarding practical or methodological challenges. An obvious downside to having an enumerator supervisor is the additional cost to the research project. There are also potential challenges for enumerators taking direction from two people; researchers need to ensure that enumerators are provided with consistent feedback on how they are collecting and coding data.

Training enumerators

Training enumerators is an often overlooked area of survey implementation. There is a tendency to want to get into the field as quickly as possible, which often results in poorly designed questionnaires, poorly trained enumerators and compromised data quality. The time required to train enumerators depends on their level of education and experience. For example, a highly educated team with some experience implementing field surveys can be trained in one week, while a less experienced team may need two weeks to understand, internalize and develop confidence in implementing the questionnaire (Whittington, 2002).

Pretesting the questionnaire with enumerators is a critical part of the research process. We recognize the chicken and egg problem inherent in pre-testing questionnaires and training enumerators: if the researcher does not speak the language that surveys will be conducted in, it is hard to pretest questionnaires unless she has at least one trained enumerator to administer the questionnaire. Ideally enumerators should be trained so that they can confidently administer the questionnaire for the pretest. The pretest process involves adjusting the questionnaire to local conditions based on information gathered during reconnaissance visits, key informant meetings and feedback by enumerators. If you are translating the questionnaire into the local language, which is generally a good idea, it is essential to make sure that all questions have retained their original meaning. Work with the enumerators on translating questions into the local language until all questions can be put across clearly and correctly. One possible exercise is to use two enumerators to do double translations: one translates from the language of the researcher to the local language, and the other independently translates from the local language back to English. If the question is understood to have the same meaning, the translation is acceptable. Questionnaires should be pretested to obtain a wide range of responses to survey questions. The objectives of pretesting are to make sure that questions are correctly understood (that words and phrases make sense for a given local context), and that options for each question fit the local context (that they adequately capture the range of possible responses). The number of pretest interviews depends on how well the questionnaire is adapted to the local context, but plan to test the questionnaire in at least 6–10 households and 1–2 villages that will not be included in your study sample (PEN, 2007).

The time it takes to answer the entire questionnaire during pretesting may seem prohibitively long and it is tempting to start shortening the questionnaire by cutting questions. The actual interview time will, however, be much shorter as enumerators gain experience with the questionnaire. It is important to remember that, in some households, sections might be skipped (for example, no member of the household has worked as an off-farm labourer). Expect the time required to administer questionnaires to drop by roughly a third as your enumerators gain experience. The researcher, field supervisors and enumerators should test the questionnaire until it is thoroughly revised, and until each enumerator thoroughly understands each question. The researcher should attend at least one pretest interview with each enumerator to be sure that the interview technique is up to the required standards and that questions are being asked in a consistent manner across enumerators.

As a general rule, pretesting should never be done in the households, villages or other units included in the study sample. In cases where only household surveys are being administered, it is alright to pretest the questionnaire with households that fall outside of the study sample but reside in the same village. However, pretesting should be done with people that resemble the target sample. Look for locations that are as close as possible to the selected research site and reflect variables of interest, such as market access, population density, agricultural potential, land tenure, ethnic diversity, forestry activities, proximity to protected areas, and so on.

Pretesting is not just about improving the questionnaire, it is also essential enumerator training. Enumerator training generally involves a variety of activities before going to the field. These include: reviewing the questionnaire(s) to make sure that all questions are commonly understood; role playing in small groups with teams of enumerators asking questions and recording responses; and role playing in larger groups with enumerators observing each other and giving feedback. Training should also include sessions on the overall objectives of the research project, research ethics, how to conduct yourself in the field, how to deal with difficult or uncomfortable situations, how to deal with problem respondents, the rationale and objectives for pretesting the questionnaires, how logistical issues related to survey implementation will be dealt with, protocols for checking questionnaires in the field, expectations about performance and a review of the project work plan. Pay enumerators for their time spent training and, if the research budget allows, train more enumerators than needed so the top candidates can be selected at the end of the training session. A suggested schedule and topics for enumerator training is described in Table 10.1. The schedule can easily be extended to two weeks by increasing the amount of time for each activity and covering fieldwork protocols that will strengthen the performance of enumerators in the field. We suggest including several group sessions that involve researcher-led lectures and discussions of various aspects of fieldwork.

Taking trained enumerators to the field and having them interview several respondents (outside the sampling frame) is a valuable opportunity for the enumerators to get acquainted with how the entire questionnaire works in practice. Pretest the questionnaire first as a large group and then later in smaller

Table 10.1 *Proposed training schedule for enumerators with prior experience*

DAY	Morning	Afternoon
1	Overview of research project (relevance, questions, hypotheses). Introduce main sections of questionnaire(s), review and clarify enumerators' understanding of rationale for main sections covered.	Section by section review of questions *(update questionnaire in response to enumerator feedback on the appropriateness and feasibility of questions).*
2	Section by section review of questions *(update questionnaire in response to enumerator feedback on the appropriateness and feasibility of questions).*	Session on research ethics (including ensuring the anonymity of respondents) and how to conduct yourself in the field.
3	Discussion of updated draft of questionnaire(s). Session on recording responses and protocols for checking questionnaires in the field.	Role playing of questionnaire in small groups *(observation and feedback by researcher, checks enumerator coding of responses).*
4	Session on dealing with uncomfortable situations or problem respondents in the field.	Role playing in larger group with enumerators observing one another and giving feedback *(researcher provides iterative feedback, checks enumerator coding of responses).*
5	Session on performance expectations and performance reviews.	Role playing in larger group with enumerators observing one another and giving feedback *(researcher provides iterative feedback on role playing, checks enumerator coding of responses).*
6	Review the work plan for fieldwork (including conducting the pretest), objectives, need for enumerator feedback, logistics. Presentation of training certificates.	

Note: Training time will need to be increased if: (a) the researcher does not speak the local language (to allow for clarification and training of the research questions in the language in which the questionnaires were created and again in the local language); and (b) the enumerators have little or no previous experience in applying socio-economic surveys.

groups, under the researcher's supervision, until enumerators feel confident about conducting interviews on their own. In particular, enumerators should become familiar with how to deal with a wide variety of responses to questions and challenges that appear when implementing them in a local context. Rehearsals are also critical to enable enumerators to conduct the interview in a less formal atmosphere. Train the enumerators to purposively vary the way the interview is conducted according to how information is conveyed by the respondent. Conducting interviews in a non-mechanical fashion puts respondents at ease and can significantly improve the quality of data collected. Anticipate several interviews before enumerators acquire these skills.

Scheduling an opportunity for enumerators to practice questionnaires in a field setting can also provide valuable insights into how enumerators will perform during fieldwork. If you hire external enumerators, they may have spent only a limited amount of time in the field and find the conditions difficult to deal with. Observe the body language of the interviewer: good enumerators show an attitude of being present and genuinely interested in the respondents' replies. They also make it clear from the onset of the interview that the information given is valued and appreciated.

Training is a gruelling process for all involved. Repeatedly going through and discussing each question at length, then making sure there is a common understanding of the question being asked requires sustained concentration by all parties. In general, if enumerators are confused by a question, then respondents will also be confused. Training should be viewed as an opportunity to gain the insights of people who may be familiar with the field setting and context, and conversant with respect to the feasibility of asking the questions you want to ask. An important part of training is to let enumerators gain their first experiences with the questionnaire through asking each other questions and by thinking through the possible responses to questions.

A final word with respect to enumerator selection is that sometimes the least educated and/or experienced enumerators turn out to be the best. Researchers should observe trainees closely during the training and early fieldwork process and remain open-minded regarding the potential for building human capital in relatively inexperienced enumerators.

Managing a field team

After hiring, training and pretesting, the research team is ready for the field. The big challenge ahead is to ensure that the enumerators remain committed to the research project and to collecting the highest quality data possible. There are several important things that the lead researcher must do: build team spirit by demonstrating commitment to the research project by spending time in the field; set clear objectives for enumerators, including how their performance will be evaluated and how good performance will be rewarded; and treating enumerators with respect. The goal is to retain the trained enumerators throughout the duration of the project, which should result in the collection of high quality data. Having to train replacement enumerators can be very disruptive to the scheduled research and may affect team dynamics resulting in poorer quality data collection.

Enumerators gauge their dedication and enthusiasm for the research project by their perception of the researcher's level of commitment. We stress the importance of spending time in the field, participating in surveys, struggling through the gruelling conditions and long days, and establishing a relationship of mutual trust and respect with the field team. Though we have no empirical basis for this statement beyond our collective experience, we think that this may be one of the most important determinants of high quality data collection. As a manager of a field research team, it is essential to communicate clearly and often with enumerators, for example, they will have questions that need to be answered, suggestions for how to improve the data collection or possibly even problems with respondents or other enumerators that need to be resolved. The simplest advice for sound communication is to be available to listen to the enumerators, and to make them feel heard and respected. Conducting regular meetings with the research team while in the field is useful for clarifying questions with the entire group. Through clear communication, both researcher and enumerators can learn from and adapt the research process to promote high quality data collection.

It is important to clarify expectations to enumerators early in the research process. The researcher should communicate expectations regarding the number of questionnaires that they should be able to complete within a given time period, and also set clear quality standards. This requires participating in interviews and paying close attention to how enumerators are recording responses. Providing constant feedback and checking questionnaires for completeness and appropriate coding of responses should occur throughout the duration of the fieldwork. If the research budget allows, consider a system of salary increases and bonuses for work well done. Enumerator salaries should be perceived as fair in the context of local market conditions. We caution against paying too high a salary as this might distort market conditions and increase expectations for enumerators who will work with future researchers. Researchers should use particular caution when employing local enumerators as high salary levels can lead to envy and conflict within communities. A good starting point is slightly above the local wage rate appropriate to the level of enumerator education. We recommend slightly higher as fieldwork often has hardships associated with it: people may be away from their families for extended periods, working under harsh and/or physically demanding conditions, and working for longer hours than the standard work day, and so on. We recommend having each enumerator sign a contract that specifies the terms of reference, salary and details regarding health insurance, days off, overtime pay, and so on.

Dealing with problem enumerators is a difficult issue. Three common problems are: enumerators that have poor communication skills and thus do not work well with respondents; enumerators that have trouble understanding some of the questions on the questionnaire and/or recording responses accurately; and enumerators that fabricate data. To detect such problems early on, consider pairing enumerators in the first week of data collection. This allows them to help each other and cross-check to develop good practices. Once things are running smoothly (and it is clear that all enumerators are competent) then they can begin to work alone. Another possibility is to let a weaker enumerator shadow a very good enumerator for a couple of days to learn about how to deal with respondents, how to ask questions and how to record responses. It is also possible to regularly schedule co-interviews where two (or more) enumerators do interviews together to ensure continued consistency. The third problem of enumerators who fabricate data and violate random sampling protocols – by, for example, interviewing households that are spatially closer than the selected households – is very tricky. Enumerators can be given a stern warning followed by a second chance. However, if it is reasonable to believe that the enumerator will not adhere to the team's data collection standards, they should quickly be replaced. The cost of finding out that an enumerator is incompetent can be unacceptably high if the discovery is made after the enumerator has interviewed a large number of households. Quick decisions can save data and money. Penalizing enumerators who violate data collection protocols also sends a clear message to other team members of the importance of adhering to standards for data collection and quality.

Finally we emphasize the importance of listening to enumerators. This is particularly important if the researcher does not speak the local language; enumerators are the conduit between the researcher and respondents and their experience is critical to data collection and data quality. Enumerators may reveal that some questions are difficult for respondents to understand, or that they feel respondents have a systematic bias in their responses. For example, if timber harvesting is illegal in the study area, enumerators may have the best sense of whether or not respondents are providing accurate information when they report figures on the quantity and value of timber harvested. Researchers should also listen carefully to enumerators as a way of assessing their morale. When enumerators get fatigued and/or overwhelmed by the workload, it is important to address concerns and take mitigative action. For example, taking them out for dinner, giving them a few days off, breaking early one day and going swimming; playing football or doing some other fun activity at the end of the day can boost morale and break up the monotony of fieldwork.

Good questionnaire management in the field

Before leaving for fieldwork, finalize questionnaires and have them printed. Think about practical issues such as how to pack and store questionnaires so that they do not get dirty or wet. This includes providing enumerators with appropriate field equipment for keeping questionnaires dry (for example, providing enumerators with plastic envelopes; zipper-locked bags; rain ponchos; and waterproof rucksacks) and minimizing the impact if questionnaires do get wet (for example, by using pencil or waterproof pens for recording). Another important issue is making sure that questionnaires will not get lost or stolen. Establish a system for keeping track of questionnaires as they get distributed to team members and have a plan for storing questionnaires in a secure place. Each questionnaire should have a unique identifier and it is a good idea to pre-number questionnaires before handing them out to team members, ensuring that duplicate questionnaire numbers do not occur.

After enumerators have conducted interviews, there is a lot of work to be done to check questionnaires. To minimize problems of missing data, incorrectly recorded responses, and so on, it is very important to devote considerable time to checking questionnaires in the field. Several systems can be implemented to check questionnaires. In Table 10.2 we present an example of a system for checking for data completeness and quality.

Detecting data problems takes time and practice. Missing or miscoded data are relatively easy to identify; more nuanced are responses that just do not seem feasible – for example, a household that has a very large livestock endowment but has very low livestock income. After spending some time with enumerators and respondents, researchers gain a better sense for which responses are feasible and what information appears odd or infeasible. In particular, researchers should check that the recorded data are consistent with other information provided in other parts of the questionnaire. This process of validation is

Table 10.2 *Protocol for checking questionnaires*

Timeline	Task
Day of interview	Enumerator conducts interview and records responses.
Evening of interview	Enumerator reviews questionnaire to make sure responses are recorded/coded correctly.
	Researcher reviews questionnaire and flags missing data; miscoded data; values appearing to be outliers; and other issues.
Day after interview	Enumerator resolves flags either by recall of interview or revisiting household.
	Researcher checks de-flagging, confirms that questionnaire is complete.
	Questionnaire is packed in dry secure place until data entry.

referred to as triangulation (see Chapter 11 for a detailed discussion of interview bias and how to deal with it). Incorporating multiple questions to elicit data for important issues helps with assessing the quality of information and allows triangulation. Evaluation of responses with the enumerator helps with detecting any inconsistency between response and recording. Finally, train enumerators to provide a response for every question. If there are blank spaces in a questionnaire, it is impossible to know if it is because the question was not relevant, the respondent refused to answer or the enumerator simply missed asking the question. It is important to have clear coding protocols for not applicable and missing responses (for example, -8 = Not applicable, -9 = Missing). All researchers should implement a system of flagging or colour-coding potential problems in questionnaires. Some researchers use small Post-it notes to 'flag' and write notes related to missing, miscoded or inconsistent data. Others write notes in the questionnaire using different coloured pens. As enumerators address each of the issues, either by recall of the interview or by revisiting the household, they can check off or respond to the queries. When the researcher reviews the questionnaire the second time, she can remove the flags or include a special mark on the coloured text, signalling that the questionnaire has been completed to the required standard of data quality.

Conclusions

Hiring and training a competent and dedicated field team is essential when undertaking complex socio-economic surveys: the team will strongly influence the quality of collected data. In assembling the field team, the decision to hire external or local enumerators is based on researcher skills, the research objective, budgetary constraints, and so on. Researcher engagement in the fieldwork at all levels is essential. Enumerators will take greater care in their data collection efforts if they find that the researcher is invested in the process. This may mean extraordinarily long days in the field as the researcher participates in interviews and checks questionnaires in the evenings. Every questionnaire needs to be carefully checked for missing values, inconsistent data, outliers, and so on, while research teams are in the field. Training is an ongoing process through researcher monitoring and communication with enumerator teams in the field. Hard work will result in higher quality data! Finally, researchers should respect enumerators and take seriously their feedback on the questionnaire, respondent reactions to questions, and so on. Enumerators are on the front line of the research and have the best knowledge regarding how respondents are interpreting the survey questions.

Key messages

- Take great care in hiring enumerators and plan for at least one week of training, translation and pretesting before starting data collection.
- Respect enumerators and work hard to maintain their commitment throughout the entire data collection period.
- Establish operational field-level protocols for data collection and management.

References

Devereux, S. and Hoddinott, J. (eds) (1992) *Fieldwork in Developing Countries*, Harvester Wheatsheaf, London

FAO (Food and Agriculture Organization of the United Nations) (1995) 'Conducting agricultural censuses and surveys', FAO Statistical Development Series No 6, FAO, Rome

Hoare, R. E. (1999) *Training Package for Enumerators of Elephant Damage*, WWF International, IUCN, Gland

NASDA (2010) 'Enumerator training modules', National Association of State Departments of Agriculture, www.nasda.org/cms/20790/7267/17757/21098/22648.aspx, accessed 23 September 2010

PEN (2007) 'PEN technical guidelines, version 4', www.cifor.cgiar.org/pen, accessed 21 September 2010

Poate, C. D. and Daplyn, P. F. (1993) *Data for Agrarian Development*, Wye Studies in Agricultural and Rural Development, Cambridge University Press, Cambridge

Puetz, D. (1993) 'Improving data quality in household surveys', in Braun, J. V. and Puetz, D. (eds) *Data Needs for Food Policy in Developing Countries*, International Food Policy Research Institute, Washington, DC, pp173–185

Ross, A. S. (1984) 'Practical problems associated with programme evaluation in third world countries', *Evaluation and Program Planning*, vol 7, no 3, pp211–218

Vemuri, M. D. (1997) 'Data collection in census: A survey of census enumerators', in Rajan, S. I. (ed) *India's Demographic Transition: A Reassessment*, MD Publications, Delhi, pp111–138

WCSRM (Web Center for Social Research Methods) (2010) 'Web Center for Social Research Methods', www.socialresearchmethods.net, accessed 23 September 2010

Whittington, D. (2002) 'Improving the performance of contingent valuation studies in developing countries', *Environmental and Resource Economics*, vol 22, no 1–2, pp323–367

Chapter 11

Getting Quality Data

Jens Friis Lund, Sheona Shackleton and Marty Luckert

Such things people cannot tell you!
Immediate reaction of a Tanzanian enumerator when presented with a socio-economic household survey questionnaire.

Introduction

Try for a moment to imagine a typical survey interview situation. A stranger walks into the house and starts asking detailed questions about the household. What is really going on in that situation? It is the making of data and it is happening in a manner and environment that is loaded with opportunities for failure! In the interview situation – typically lasting between 30 minutes and two hours – the respondent and enumerator should gain a common and hopefully correct understanding of the real-world experiences that the enumerator asks the respondent to communicate. This brief exchange of information takes place between two or more people that normally are strangers, implying that different pre-understandings of concepts and reality, as well as potential mistrust, should be overcome within a very short time. The enumerator is normally temporarily employed, with no personal interest in the research or the data, other than possibly wanting to minimize the time spent on his work (the interview). In addition, the enumerator must correctly transfer the responses into the pre-designed format of the questionnaire, while keeping the interview flowing. The respondent may want to present him or herself in a favourable manner. If the respondent believes the research is potentially influential, he or she may also want to answer strategically, for example, to complain about politics or economic conditions. Furthermore, the required information will often be regarding sensitive issues such as economic assets, diseases or illegal incomes.

In short, the chances that one will get quality data would from the outset seem rather slim! Unfortunately, the literature on empirical household studies rarely provides detailed information on how research is implemented: in other words, how enumerators were trained; whether and how triangulation was done; and what criteria were used to evaluate data quality. This chapter addresses the issue of how to pursue data quality through attempting to avoid systematic measurement errors, in other words, errors arising during the implementation of the survey that systematically affect the measurement of a variable across a sample (see Box 11.1). A systematic error creates either positive or negative bias in the sample estimate of a variable. The tendency for people to under-report on income from illegal activities is an example of systematic measurement error, and we shall look at how to minimize the risk of recording such errors. This involves a discussion of factors surrounding the interview situation, the enumerators and the nature of the relationship between researcher and respondents. We focus only on systematic measurement error and on the data

Box 11.1 *On measurement error*

Any estimate from a sample (for example, gross income from maize) may be thought of as having two components: the true value plus a measurement error. The measurement error component can be further divided into two parts: random and systematic measurement errors.

Random measurement error is caused by factors that randomly affect the variable estimate across the sample. For example, for an estimate of gross income from maize, random measurement error could arise from the tendency of respondents to include/exclude maize harvested at a point in time before/within the recollection period of three months. Importantly, random error does not have any consistent effect across a sample. The randomly distributed positive and negative errors will sum to zero across the sample. Thus, random error adds variation but does not affect the average value of the estimate.

Systematic measurement error, on the other hand, is caused by factors that systematically affect the variable estimate across the sample. With our example of gross income from maize, respondents may answer strategically and under-report their harvest if they perceive that by looking poor they may attract some external assistance. This can also be an effect induced by some enumerators who, unknowingly, give respondents an impression that appearing poor would be beneficial. In both examples, the error tends to consistently influence negatively the average value of the estimate.

Source: Based on Trochim (2006)

collection situation, as other parts of the research process affecting data quality are covered in other chapters of this book, including sampling (Chapter 4), collecting contextual information (Chapter 5), questionnaire design (Chapter 7), implementing the survey (Chapter 10), and data entry and quality checking (Chapter 12).

Systematic measurement errors

This section outlines a number of underlying reasons for systematic measurement errors that may lead to bias in data. We also provide guidelines on how to detect and avoid or minimize them. We categorize these reasons according to whether they originate with the enumerators and/or questionnaire, the respondents' strategic behaviour and understanding, or bounded knowledge.

Enumerators and questionnaire administration

Personal characteristics and appearances of the enumerator

The problem: The age, sex, ethnic group, caste, attitude and appearance of enumerators can greatly influence the data generated in a survey. An important choice is whether or not to use local enumerators (see also Chapter 10). A major advantage of using local enumerators is their knowledge and familiarity with the area, language and livelihood strategies of the respondents. Local enumerators may also have lower salary expectations and reduce the need for extra lodging and transport. A disadvantage may be that finding experienced and/or sufficiently educated enumerators is difficult, hence triggering potentially higher demands for training and supervision. In particular, enumerators must be able to learn and apply new procedures consistently. Moreover, the familiarity of local enumerators may be a drawback when the research touches upon sensitive issues (for example, incomes, diseases, illegal activities). Many respondents may be more comfortable providing such information to strangers who will leave the area, rather than locals whose lack of confidentiality might have a high social cost to the respondent.

Another important characteristic is the social skills of the enumerator. Some enumerators have great interpersonal skills and can immediately create trust with people they meet. This is not something that can be easily picked up through training, and therefore becomes an important selection criterion. Interpersonal skills can be observed through interaction with enumerators and with respondents in the field. Regarding gender, women may in some cases only share knowledge about medicines for women's health with female enumerators.

Similarly, female enumerators may be best for interviews about domestic affairs, child care or craft production. Male enumerators may be better at asking questions related to typical male activities, such as hunting or timber extraction.

Avoiding/minimizing the bias: Being picky in selecting enumerators is a key success factor. Make sure you see how your candidate enumerators work in the field before making hiring decisions. Some researchers prefer training more enumerators than they need, having informed the trainees in advance that only a subset of them will eventually be hired. Key attributes to consider include: Are they trustworthy? Can they complete the questionnaire in reasonable time without tiring the respondents? Do they understand and follow your directions? Through careful selection and constant encouragement and supervision you can help to build these attributes in your enumerators.

Integrity of enumerators

The problem: Unfortunately, some enumerators tend to take short cuts or have unacceptably low quality standards in survey implementation or even go to the extreme of falsifying data. Detecting such misbehaviour can be difficult, but it may be useful to consider the motives. The motives may include low motivation, confusion about the overall purpose of the survey, difficulties in getting to talk to the household, fear of engaging people with sensitive questions or the desire to please the researcher with interesting results.

Avoiding/minimizing the bias: The first line of defence against careless data collection and falsification is prevention. Workloads of enumerators need to be realistic. Enumerators need to be well-trained to be comfortable asking all the questions and to understand that the researcher is not interested in embellished data. Doing spot checks with households to see whether the enumerator came to ask questions, and letting enumerators know that these checks will occur, may prevent the problem. It may help to explain to enumerators that there are consistency checks built into the survey that will reveal if the data has been fabricated or collected carelessly.

The second line of defence is *ex post* detection. Keeping careful track of data as they are collected is essential, for example, a quick review of questionnaires every evening and follow-up meetings with enumerators to go over perceived 'odd' data (see Table 10.2 in Chapter 10). In some cases, enumerators will have to return to the household to verify dubious data. In short, be suspicious of data appearing too similar or too random. Moreover, systematic differences between results obtained from enumerators can, at the extreme, be an indicator of careless collection or falsification. Yet, differences in the way enumerators understand and pose questions (as well as probing) can also trigger differences. The issue should, therefore, be investigated thoroughly before accusations are made.

Researchers should strive to be present at as many of the survey interviews as possible to get a sense of how the data is being collected. This knowledge will not only help in detecting careless collection and falsification, but can also help identifying other problems and provide opportunities to offer encouragement and recognition for a job well done. If working with a large enumerator team, checks of data quality can be achieved through nominated group leaders with systematized checking of the questionnaires. Finally, creating a good working environment will minimize the risk of such problems occurring. Bonuses, honours, praise and individual care can all contribute positively.

Enumerator fatigue

The problem: Conducting the same interviews day in and day out can be tedious and monotonous. Enumerators may quickly become demotivated and demoralized.

Avoiding/minimizing the bias: Enumerators need constant recognition, support, feedback and motivation for them to give their best. For intensive large-scale or repeat surveys, it is essential to build in activities keeping enumerators interested and enthusiastic. This may include knowledge sharing, feedback meetings highlighting interesting findings, training identified by the enumerators and social activities. One of the authors of this chapter coordinated a demographic and health census of 10,000 households, undertaken by only ten enumerators who worked on this for almost a year. Most of her energy and time was spent keeping the enumerators motivated through regular field visits, accompanying them to households, recognizing their worth by asking them for their opinions and ideas on the process and on the data, providing training, organizing social get-togethers, sharing key and exciting results, and so on.

Probing bias

The problem: Enumerators should not be trained to act like field robots – they will then not be able to develop a rapport with households. But differences between enumerators regarding interpreting, explaining and exemplifying questions may influence the answers. Imagine that a questionnaire seeks to elicit households' forest income over the past year. One enumerator might do this by asking several questions about the household members' daily activities and probe using various examples of forest products that he knows are commonly used in the area – thereby facilitating the respondents' memorizing. Another enumerator might simply state the question once and record whatever the respondent answers without asking any follow-up questions. Clearly the former enumerator will end up with a higher average forest income estimate than the latter.

Avoiding/minimizing the bias: All enumerators should, as far as possible, use the same approach to probing. For example, prompting for open-ended questions should be standardized. The risks associated with probing biases increase if questions are stated in general or abstract ways, which often require further explanation (see Chapter 7). During pretesting of questionnaires, such sources of bias can be detected, for example, by role playing the questionnaire and by regularly supervising how enumerators ask and exemplify the questions. Preventing such sources of bias is best done by spelling out all questions in their entirety, having detailed written guidelines for the questionnaire, and thoroughly training the enumerator team.

Bias arising from choice of respondent

The problem: Households are many-headed creatures and one head may not know what the other is doing. As an example, in many rural areas of developing countries women and children fetch firewood and water, while business and hunting are the domains of men. The division of labour and responsibility in households, however, varies from place to place implying that no rules of thumb can be given. One should be aware that the choice of respondent will in all likelihood influence the answers.

Avoiding/minimizing the bias: There will clearly be a trade-off between time needed to do the survey and the opportunities for getting fuller pictures of household livelihoods. If several household members participate in answering the questionnaire, they can respond to different questions. For example, children may be asked about products that their parents are not likely to know much about, such as collecting fruits and hunting small birds and animals. But dominant family members could object to letting those answer who are best suited. Because of such problems, resources permitting, interviewing individually one adult male, one adult female and one child can give a more detailed picture. No matter what approach is chosen, it is important to document in the sampling procedures who was answering the interviews.

Respondents' strategic behaviour and understanding

Even if enumerators are full of integrity, well-trained and do an excellent job in asking questions, some biases may arise due to respondents.

Strategic answering

The problem: Strategic answering arises as a consequence of push and pull factors. Foddy (1993) uses the term 'threat factor' and provides three overall categories: idiosyncratic, social rejection/acceptance and sanctions/rewards.

First, idiosyncratic questions focus on issues that generate personal anxieties with particular respondents. For example, someone who has recently been robbed may choose to under-report household wealth because of fears that disclosing asset information could increase robbery risks. Or some households may be hesitant to reveal their use of medicinal plants out of fear that it may be frowned upon by the church to which they belong.

Second, questions may also generate fear of social rejection or, conversely, hope of social acceptance. For example, questions regarding sexual behaviour, sexually transmitted diseases, illegal harvesting, behaviour that is likely to cause ridicule, or large dependence on relatives (remittances) can all be seen as potential threats to respondent social status and should thus be posed with care.

Third, questions may also generate fear of political or economic sanctions or, conversely, hope of rewards. Questions regarding illegal activities, political activism or altruistic behaviour can all trigger expectations about sanctions or rewards. Questions related to wealth and income may be sensitive for various reasons. In many settings, being perceived as wealthy carries with it obligations and responsibilities: being generous, providing labour opportunities and relief support to the less well-off. Overstating poverty and understating wealth may also result from hope that those using the research outputs will be more likely to help respondents, for example, by initiating a development project such as building a school. The sheer presence of and impression made by researchers in a rural area can generate expectations of development aid or other rewards, which may result in people trying to appear poor and worthy of assistance.

Finally, the history of a field site may provide important hints as to what issues are contentious. If working with natural resource issues, for instance, sites with previous presence of conservation projects or non-governmental organizations (NGOs) will often display a heightened awareness of what is legal and what is not, and, hence, increase the risk of strategic answers.

Avoiding/minimizing the bias: For all three threat categories mentioned above, strong feelings and strategic answers may be generated even by seemingly harmless questions. In cases where these threats are present, a number of preventive measures may be taken. First, it is of utmost importance to avoid the spreading of false rumours by providing ample information regarding research purposes and to make sure this information is disseminated widely in the community through appropriate channels, such as community meetings, local radio stations or an information brochures in the local language. Second, in every individual interview, this information should be repeated, possibly including background information to sensitive questions. Also the promise of confidentiality should be made very clear to the respondents at the beginning of the interview. In some cases it may not be necessary to record respondents' full names – at least not while conducting the actual interview. This may make people more comfortable.

To avoid strategic answering, it is also important that the answer options should contain the full range of potential responses, including, for example, illegal activities. A key factor is the level of trust between the enumerator and the respondent. This trust can be strengthened through approaches that make it clear that answers will be held in strict confidence. Typically, in repeated surveys it proves easier to get honest responses to sensitive questions during subsequent household visits, as trust is being built over time. Spending significant time in the village and participating in social activities, for example, in the Friday night local football game or local party are all measures that will increase trust and diminish the likelihood of strategic responses.

Normally the respondent's homestead is chosen as the place to conduct the interview. This may, in addition to facilitating trust, yield the benefit of triangulating household responses, through visible characteristics, such as the telling evidence of a hunting trap lying around in the courtyard or the expensive newly bought furniture in the living room. Using such observations should, of course, be done in a polite, or perhaps humorous, way in order to build a relationship of trust. This also diminishes the probability of strategic biases. If the respondent seeks to avoid doing the interview at the homestead, it could be an indication that something is being hidden.

Another way to check on at least some of the information provided by respondents is to use built-in triangulation in the questionnaire itself. This can be particularly rewarding in research that involves multiple survey rounds where information obtained in previous rounds can be used for validation purposes in later rounds. Box 11.2 provides some examples.

The temptation of the respondent to provide misleading answers may also be diminished if the enumerator demonstrates good background knowledge of the topics and local conditions, including sensitive ones. For example, in inquiring about illegal activities, it is important to have a good knowledge about the level of enforcement of laws and regulations in the communities (Shackleton et al, 2001, pp134–135). In one survey in Tanzania, enumerators often experienced that respondents first explained to have produced 'only two small bags of charcoal'. Normally, however, charcoal is produced in kilns that yield much higher quantities. By revealing this knowledge with a knowing smile, the enumerators made the respondents laugh and reveal the true amounts.

Box 11.2 *Triangulation using information previously obtained in a survey*

It is possible for researchers to use data collected earlier in a survey, to increase the quality of ongoing data collection. For instance:

- Household assets may point towards specific types of income sources. Having recorded ownership of items such as scotch carts, ploughs, oxen, shops and various tools should lead the enumerator to check carefully the income derived from activities related to such assets.
- Household member skills and general livelihood strategies should be elicited in the first survey round to provide a checklist for later survey rounds. Knowing *ex ante* that a household member is a carpenter, house builder, charcoal producer, bike repairman, herbalist, and so on, is valuable in later survey rounds to assure that the income is elicited.
- The household food security situation – food storage – is a strong predictor of future activities to supplement income. Hence, knowing how many months the main staple crop lasts for the household's own consumption is valuable information.
- Household religious beliefs may be related to an unwillingness to reveal certain types of information, for example, the use of traditional medicine or income from certain types of businesses. In such cases, enumerators should be more careful and critical about this type of information.
- Asking households about cash costs and own consumption of various items, for example, agricultural produce, can be a good way of triangulating information on savings, cash income and agricultural storage and harvest.

In general, the information obtained in previous survey rounds should provide some information about data that will be obtained in later rounds. Hence, enumerators should be suspicious of households that suddenly report very different incomes, if the change cannot be explained by seasonality or some other cause.

Finally, the 'threat' level of sensitive questions can be reduced by various techniques in the way the question is posed and the answer options presented. Box 11.3 provides a few examples and references.

Respondent learning in panel studies

The problem: If an enumerator visits a household more than once, then respondents may think that they already know the questions, and proceed to answer them 'automatically', without carefully listening to and thinking about the detailed questions.

Box 11.3 *How to ask about sensitive issues*

Below are five approaches to asking the sensitive question: 'Did you do any illegal hunting last year?'

Approach	Question
Casual approach	Did you happen to do any illegal hunting last year?
Give a numbered card	Would you please read off the number on this card that corresponds to what you did [hand card to respondent]: (1) I did do some illegal hunting last year (2) I did not do any illegal hunting last year (3) I did hunt last year but am unsure if it was illegal or not? (4) I did only legal hunting last year (5) I did not hunt last year
The 'everybody' approach	As you know, many people do illegal hunting in this area. Did you happen to do any illegal hunting last year?
The 'other people' approach	(a) Do you know any people who have done illegal hunting? (b) How about yourself?
The 'sealed ballot' technique	We respect your right to anonymity. Please complete this form, indicating whether or not you have done any illegal hunting last year, seal it in the envelope and place it in the box marked 'Secret Ballot'.

Source: Adapted from Gray (2009, p347)

Note that the last of the approaches, the 'sealed ballot' technique, does not permit one to assign illegal hunting to individuals, but only to infer levels among a population of individuals. Another approach to infer levels of sensitive practices among a population is the Randomized Response Technique (RRT) (Gavin et al, 2009). The RRT uses a randomizing activity (such as a coin toss) and asks respondents to remember, but not disclose, the result. Respondents then select one of two undisclosed questions: one sensitive (for example, 'did you do illegal hunting last year?') and one asking about the result of the randomizing activity (for example, 'did the coin toss give "tails"?'). Respondents answer only yes or no to the question. The interviewer does not know which of the questions respondents answer, thereby providing a veil that removes disincentives for strategic answers. The interviewer does know, however, the probability that the respondent answers the sensitive question (for example, half the respondents answer the question on illegal hunting) as well as the probability of a yes/no response to the other question (half get 'tails' in the coin toss). Thereby, aggregate estimates of illegal behaviour can be obtained.

Avoiding/minimizing the bias: Explaining why the data is being collected repeatedly can help avoiding mechanical responses. For example, explaining that seasonal differences in income and expenditures are a research focus may raise respondent alertness regarding seasonal changes, and thus ensure they treat this interview as being separate from previous ones. Thus an incident from a repeat survey in northern Ethiopia can be avoided: one respondent was very apologetic for answering wrongly last time – what other reason could the enumerator have to come back and ask the same questions again?

Vague/imprecise responses

The problem: Imprecise responses are common in detailed household income accounting studies. Consider the question: 'How often did you collect medicinal plants in the past three months?' The following are three possible responses: (a) 'It depends'; (b) 'I collect when someone in the family is sick'; and (c) 'I usually mainly collect in the rainy season'. None of these responses is likely to give you the exact information that you are seeking.

Avoiding/minimizing the bias: One solution could be to take another look at the way the question is posed in the questionnaire – can it be put more precisely? Another is to make sure the enumerators fully understand what data are needed. This sequence will likely require the enumerator to break down the questions into smaller, more specific pieces. For the first response these follow-up questions could provide the information needed: *Enumerator:* 'What does it depend on?' *Respondent:* 'It depends on when we run out.' *Enumerator:* 'How often over the past three months have you run out of and gone to collect medicine?' *Respondent:* 'Two times.'

Misunderstanding the question

The problem: Respondents may easily misunderstand or misinterpret questions, for various reasons. First, long questions with complex wording run a great risk of being misunderstood (Foddy, 1993). Although misunderstanding may be prevented by using examples or illustrations, respondents may then focus excessively on the latter, rather than the general concept. Second, if the context or purpose of the question is unclear to the respondents, they will likely make their own, often wrong, guess of what information the researcher seeks.

Avoiding/minimizing the bias: Keeping questions short and concise, and avoiding difficult wording, is essential (see also Chapter 7). If questions are clear, then examples may not be necessary. When used, examples should be directly related to the question. For example, if you are trying to elicit expenditures made within the last three months, potential expenditure categories (for example, school costs) may be used as examples that can also provide a systematic probing structure within open-ended questions.

Respondent fatigue

The problem: Tired respondents may give 'quick' answers in order to finish fast, especially in the latter part of the questionnaire. This behaviour may happen as a consequence of poor timing of the interview (the respondent is tired or has other pressing matters) or because the questionnaire is too long. It may also abound in sites where a lot of surveys have already been done (research fatigue). The result will often be biased answers, because the respondent will tend to answer 'no' in order to go faster through the questions.

Avoiding/minimizing the bias: First, selecting study areas that are not 'over-researched' provides a context that is less exposed to research fatigue. Second, interview schedules should suit the respondents, vis-à-vis their daily routines. Third, in the case of long questionnaires, it is worth adjusting research plans and budgets to allow for interviews to be interrupted for later completion, whether after a short break (for example, offering a cold drink, tea or biscuits) to recharge the batteries of tired respondents and enumerators, or on another day whenever the respondent has more time available.

Bounded knowledge

When designing questionnaires, it is important to remember that human beings may fail to know or recall all the information that interests a researcher.

Recall

The problem: One form of bounded knowledge concerns recall, in other words, respondents' cognitive abilities to fully remember past events and activities. Several studies have documented that quantitative data on income and labour based on long recall may underestimate actual levels (for example, Gram, 2001). A problem often referred to when discussing recall is 'telescoping bias' that occurs when respondents draw/push past events forward/backward in time in relation to when they actually occurred, thereby creating a bias.

Avoiding/minimizing the bias: In general, questionnaire recall periods should match respondents' recall abilities. Shorter recall periods are usually associated with small and frequent events where extrapolation to a longer period is reasonable, whereas longer recall periods typically are associated with infrequent, large or rare events that may be missed out with short recall periods. The problem of telescoping biases may be eased by relating timing to significant events in respondents' lives, for example, did it happen before marriage, before the local bridge was completed or before the re-election of the president? Past experience regarding recall in other surveys and thorough pretesting can help fine-tuning questions to respondents' recollection ability.

Finally, adequate timing of surveys can help, for example, starting just after the crop-harvesting season will be good if agriculture is the main focus. For environmental incomes, with many extractive products following distinct harvesting cycles, optimizing the timing may be more difficult. Repeated surveys covering a full calendar year might then be preferable.

Differences in perceptions/understandings of definitions

The problem: When designing questionnaires, key definitions cannot be taken for granted. For example, what is a 'forest'? To European citizens, it might mostly be managed plantations with few tree species, to an Amazon dweller it may be a place with tall trees forming a naturally dense crown cover and a diverse understorey, while respondents from the semi-alpine Himalayas may refer to low hillside shrubs. Since responses depend on perceptions, having a common understanding of key words is crucial.

Differing local units also have to be standardized to allow quantities to be properly compared and aggregated. For example, a bucket may hold 10 litres in one village and 20 litres in another, while a barrel may hold 50 litres in both villages. Any comparison between these two villages requires standardization to litres.

Avoiding/minimizing the bias: Being aware of different meanings, units and perceptions is an important step in avoiding such biases. In principle, one must define all key terminology, and ensure that enumerators know and actually apply the terminology consistently when explaining the words to respondents.

Standardization of local units such as handful, headload, bundle, bucket, and so on, is a time-consuming but necessary work component. This may require a separate survey, if there is much variability within the locally applied units. For example, if there are different sized buckets used, then you may have to obtain sizes from a sample of buckets to derive a mean value. It may be necessary to collect data regarding standardization at the community level, so that differences in units between communities are detected. Often price information should be included, so that commodity unit values can be properly compared (see also Chapter 8).

Conclusions

Getting quality data from household surveys in developing countries is challenging. Researchers often operate in unknown territory with only scant knowledge regarding livelihood strategies and practices. In addition, the use of enumerators adds another source of potential measurement error.

In this chapter, we have sought to provide some operational criteria for assessing the quality of livelihoods data and some advice on how to avoid systematic measurement error. A workable approach is to minimize the potential biases arising from systematic measurement error that can distort information from being representative of the truth, and to attempt to detect the direction and severity of remaining biases through replicating results, frequently using alternative approaches. This approach implies a belief that we will obtain high quality data by seeking to avoid biases and to triangulate our data from multiple sources. It also implies that we should be sceptical regarding the quality of data that are not supported by many and good reasons for their truthfulness, for example, that: defy commonly accepted theories; are challenged by other sources; do not fit the understanding of the local context; may be biased by social identity and power relations; and rest exclusively on one single source and cannot be supported by other evidence. But data that are characterized by these features should not be dismissed out of hand. If we rejected all data that defy common theory or our previous understanding of the local contexts, we would also fail to reject wrong theories and to correct wrong understandings of local contexts. In other words, the search for biases should not blind researchers' curiosity, nor jeopardize their ability to challenge common wisdom – arguably, this being one of the noblest functions of science.

Key messages

Several of the potential solutions cut across various of the identified biases. We summarize them in five key recommendations. Giving careful thought and adequate time and attention to these five recommendations will go a long way to minimize the biases discussed, and to ensure that your data – and the research results and conclusions they generate – hopefully can be trusted.

- Take care in selecting enumerators with an eye to their integrity, interpersonal skills and ability to pick up and apply consistently new knowledge, and train them carefully.
- Understand how the local social, cultural and political context may influence peoples' willingness to share desired information.
- Spend time in study sites and participate in interviews to check enumerator performance and consistency, and to get a feel for the reality behind the data.
- Triangulate data from multiple sources.
- Create and implement standard procedures for data quality checking while in the field.

References

Foddy, W. (1993) *Constructing Questions for Interviews and Questionnaires: Theory and Practice in Social Research*, Cambridge University Press, Cambridge

Gavin, M. C., Solomon, J. N. and Blank, S. G. (2009) 'Measuring and monitoring illegal use of natural resources', *Conservation Biology*, vol 24, no 1, pp89–100

Gram, S. (2001) 'Economic evaluation of special forest products: An assessment of methodological shortcomings', *Ecological Economics*, vol 36, pp109–117

Gray, D. E. (2009) *Doing Research in the Real World*, Sage, London

Shackleton, S. E., Shackleton, C. M., Netshiluvhi, T. R., Geach, B. S., Balance, A. and Fairbanks, D. H. K. (2001) 'Use patterns and value of savanna resources in three rural villages in South Africa', *Economic Botany*, vol 56, pp130–146

Trochim, W. M. K. (2006) 'Research Methods Knowledge Base', www.socialresearchmethods.net/kb/index.php, accessed 31 August 2010

Chapter 12

Data Entry and Quality Checking

Ronnie Babigumira

At one level 'data' are the world that we want to explain ... at the other level, they are the source of all our troubles.
Zvi Griliches (1985, 'Data and econometricians: The uneasy alliance', *American Economic Review*, vol 75, no 2, pp196–200)

The ethnic theory of plane (and data) crashes

In his compelling book *Outliers: The Story of Success*, Malcolm Gladwell presents what he calls the ethnic theory of plane crashes: '[P]lane crashes rarely happen in real life the same way they happen in the movies. Some engine part does not explode in a fiery bang. The rudder doesn't suddenly snap under the force of takeoff... Plane crashes are much more likely to be the result of an accumulation of minor difficulties and seemingly trivial malfunctions' (Gladwell, 2008, p183).

This is true for data quality problems. Rarely does the 'computer' corrupt data, although a computer virus may do so or even delete data. Instead, small mistakes at various stages can seriously compromise data quality. The research question, sample selection, questionnaire design, enumerator training and survey implementation, all of which have been discussed in preceding chapters, impact data quality. However, there are computer specific sources of errors that impact data quality, the most critical being data entry and storage. A poor data entry system is a gateway to an unnecessarily large number of data entry errors, while failing to organize and archive data in a way that they can readily be retrieved and analysed can render them unusable.

The purpose of this chapter is to look at these computer-related data problems. Drawing on the Center for International Forestry Research's (CIFOR) experience with the Poverty Environment Network (PEN), this chapter identifies potential sources of errors and makes some suggestions on ways in which these errors can be prevented or minimized. Recognizing that some errors may still slip through and that some accurately transcribed values may have been recorded in error, approaches to data validation and checking are also discussed.

Because, as will be argued later, working with software one is familiar with tends to minimize errors, the ideas presented in this chapter will largely be software independent. However, for some illustrations, Stata (StataCorp, 2009) – the program used for PEN data management – will be used. Also, the following assumptions are made: (a) that data will be primary data collected using a questionnaire, such as a household survey; (b) that these data are mostly numerical, a mix of quantitative and qualitative information with the later being assigned numbers or codes for data entry; and lastly (c) it is assumed that text-heavy sections of the survey such as notes made during or at the end of survey will simply be compiled into some sort of narrative and are not checked for quality.

The chapter proceeds as follows. First it gives a brief definition and examination of the concepts of data and data quality as they will be used in the chapter. Second, it examines the building blocks critical for ensuring data quality – namely coding, data entry and data validation/checking. Third, the chapter ends with a summary and, through the conclusion, synthesizes the discussion into a few take-home messages.

Improving data quality

Data and information are often used interchangeably yet they are distinct (Audit Commission, 2007). In this chapter, *data* will refer to numbers and words that are yet to be organized or analysed to answer a specific question. These are sometimes referred to as raw data, in other words, the numbers and text recorded on the questionnaire and later transcribed to the computer.

Data quality has many dimensions, however, most of these can be distilled into the simple yet comprehensive definition: 'Data are of high quality if they are fit for their intended uses in operations, decision making, and planning. Data are fit for use if they are free from defects and possess desired features' (Redman, 2001, p73). This fitness for use comprises two characteristics; possession of desired features and freedom from defects.

The desired features include relevance, comprehensiveness, proper level of detail, ease of reading and ease of interpretation, while data are free from defects if they are accessible, accurate, timely, complete and consistent (Redman, 2001). These features have been, implicitly or explicitly, addressed in other chapters of the book. For example, timeliness of research is covered in Chapter 3, while relevance, comprehensiveness and the proper detail are covered in both Chapters 3 and 4. Ease of reading and interpretation, which arguably has more to do with information (processed data), is addressed in Chapter 14.

Our focus is on data defects in terms of accuracy, completeness and consistency. A comprehensive treatment of the each of these concepts can be found in, among others, Lee et al (2006) and Refaat (2007). The section below is largely drawn from these two references.

Accuracy: This is perhaps the most likely aspect of data quality that comes to mind when you ask scientists about data quality, and is often used to mean that data are free from errors. Refaat (2007) defines errors as an impossible or unacceptable value in a data field. Errors can be typographical ('mlae' instead of 'male'), mathematical (income not equal to gross value less purchased inputs) or related to measurement (such as the location, area, quantities).

Completeness: Data completeness refers the degree to which data are missing. According to Lee et al (2006), it can be viewed from at least three perspectives: schema completeness, column completeness, and population completeness. Schema completeness is the degree to which entities are not missing from the schema. Attrition is a good example, where, for example, households that were originally part of the survey drop out in later rounds. Column completeness refers to the degree to which values are missing in a column of a table. This is the most common type of completeness problem in socio-economic surveys and the subsequent discussion on missing data in this chapter will mainly be about column completeness. Lastly, population completeness refers to the degree to which members of the population that should be present are not. This is a usually a sampling problem.

Consistency: Data consistency may be defined in terms of consistent values among different fields, as well as values of the same field over time (Refaat, 2007). For example, children should not be older than their biological parents. The area under agricultural production cannot be greater than the total area owned or operated by a household. It is common to have questions whose responses depend on earlier responses. For example, the response to the question on how severe the crisis was depends on the response to the question of whether or not the respondent faced the crisis. Questions like these are added to improved data quality and are good candidates for consistency checks.

Data entry and quality checking: Work flow

Picture the start of fieldwork. The study site has been selected, the sampling frame determined and the sample drawn. Enumerators have been trained and the questionnaire has been pretested, improved and is now ready for use. A data entry system is in place as is a basic codebook. The survey kicks off and questionnaires begin to come in. What next?

Trying to work efficiently and, given that it is technically feasible (you can use a computer), it may be very tempting to hit the proverbial n birds with one stone: start entering the data and, in the process, scan the questionnaire for errors while also assigning codes for categorical data. This may create the illusion of saving time; however, doing this is not only inefficient – it is a recipe for errors.

It is inefficient because of interruptions that will, more often than not, arise because of illegible answers, gaps, and other inconsistencies, which will require that the questionnaire be returned to the enumerator for correction or even to revisit the household. The challenge is how to deal with the interruption. Do you pause data entry for that questionnaire until the errors have been addressed, or, do you discard what you have entered and start all over again? Pause and resume and you may enter the same data more than once, discard and you will have wasted precious time.

A more efficient and less error-prone way to work is to think of data entry and checking as a process consisting of four tasks: visual inspection of the survey instrument, coding, data entry and data checking. These tasks are best performed sequentially in a way that is akin to the flow chart of Figure 12.1.

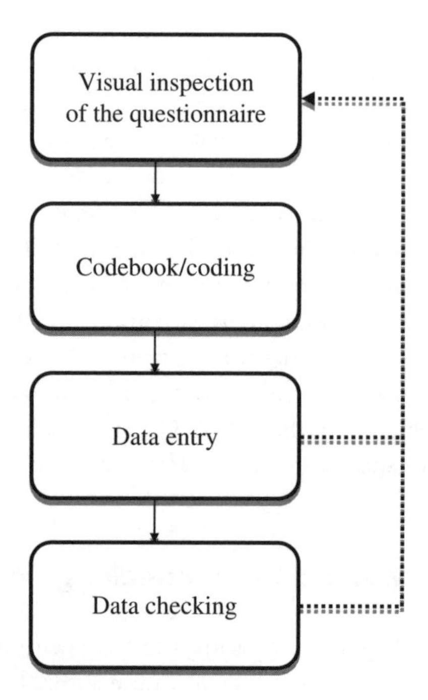

Figure 12.1 *The process of data checking*

Visual inspection

The goal of visual inspection is to make sure that all sections of the questionnaire that should be filled in have been filled in. Check that all standatd information, that is, information that does not depend on the respondents (such as location, date of interview, household code and so on) is filled in.

Visual inspection is also the first step in ensuring that data are usable. It is not enough that the questionnaire has been filled in, the recorded responses should be clear and therefore make the questionnaire easy to transcribe to the computer.

Visual inspection is best done in the field because of the ease of making follow-up visits for clarification or collecting missing information if neccessary. If you are working with many enumerators, then it is recommended that you have a debriefing session every evening where enumerators go through the questionnaires for these issues and share their experiences. This is a great learning opportunity, especially if one cannot be present during the interviews during the day.

Coding

Most data entry programs will accept both numbers and text; however, it is better to enter numbers. To illustrate, consider the hypothetical data set in Table 12.1 where, for one household, John Mayne, we have data on the main crop on his four plots. The variables are the name of the household head (**name**), plot number (**plot**), the sex of the plot manager (**sex_pm**), and the main crop grown on the plot (**main_crop**).

Looking at Table 12.1, one may immediately be drawn to the errors in the spelling of the name. While these have been exaggerated for illustration purposes, such mistakes are not uncommon especially when you have more than one person entering data. *John Mane* may stand out as an obvious mistake, however, as far as the computer is concerned, *Mayne John,* which reverses the name order is a different person as is *john Mayne*, which only differs because of the lower case in the name *John.*

Table 12.1 *Hypothetical data to illustrate problem with using text rather than numbers*

Name	*quarter*	*sex_pm*	*main_crop*
John Mayne	1	Male	Maize
john Manye	2	male	Cassava
Mayne John	3	Male	Bananas
John Mane	4	m ale	Potatoes

Table 12.2 *Further illustration of the problem with using text rather than numbers when coding*

Sex of plot manager	No.	Col %	Cum %
Male	1.0	25.0	25.0
m ale	1.0	25.0	50.0
Male	1.0	25.0	75.0
male	1.0	25.0	100.0
Total	4.0	100.0	

More interesting, however, is the variable 'sex_pm' where it appears that the only offending observation should be in the fourth row where there is an obvious gap between *m* and *a*, and yet, when we look at the data, we find that each entry of 'male' has been treated as a different category, see Table 12.2.

The reason all four have been treated as different is because of what is called 'air' or leading/trailing blanks. These are often non-perceptible spaces created when you hit the space bar before or after data. Regardless of where the air is, the effect is the same, it changes the entry. For a computer to consider two bits of text as the same, they have to be identical. This includes any leading or trailing blanks as well as the case of the letters, hence the treatment of *male* and *Male* as different.

These issues can be addressed: a variable can be trimmed to remove leading and trailing blanks cases of all observations can be standardized (all changed to lower). However, this creates additional work and is a potential source of errors. As a preferred alternative, data can be entered as numbers.

Advantage of using codes

Coding is the process of assigning numbers to text and the document containing all such codes as well as variable names and descriptions is known as a codebook or coding frame. Using codes instead of text has several advantages, which are summarized below:

Error and ambiguity: Numbers are less error prone than text. One can enter 10 instead of 1, but that pales in comparison to the errors one can make when entering text. There is also generally no ambiguity with numbers, 1 is 1 but, as shown above, small changes in text introduce ambiguity that impacts data quality. Are Cassava, cassava and cassavva different crops? In the example above, if we had in addition to the household name assigned a numeric household code to each household, we would not have had to worry about spelling mistakes in John Mayne's name. The same problem would be addressed if, instead of entering the sex variable as male/female, we used 0/1. Lastly, numbers are not case-sensitive, and tend not to have the 'air' problem common with text.

Restrictions: Most data entry programs allow users to restrict entries in a given column to a range of values (for example, between 1 and 10) or a list (for example, 1, 2, 3). These restrictions, also known as validation rules, would prevent the entry of illegal values unless they are changed. It is harder to set validation rules with text even though you could for example restrict the length of the text entered or create a drop-down list with pre-entered text from which a user can select a response. While these can help, it remains much easier to work with numbers.

Ready to use: Data entered numerically are ready to use at subsequent stages. In the example above, entering male/female for the sex variable would require that the variable is converted to a numeric variable before using it in, say, a regression. Related to this point, it is advisable, when designing the questionnaire, to try and pre-code as much as possible. It saves time and costs of data handling and also one source of error: error introduced at the coding stage (Swift, 2006).

Special issues related to coding

Missing data: Missing data is probably one of the biggest data quality problems and, depending on the fraction or patterns of missing cases, can render a data set unusable. Data can be missing for a number of reasons, what is important is that the end-user knows why data are missing and this requires that one is able to distinguish between different types of missing data.

As an illustration, consider the fictional data in Table 12.3 below, with the variables **ownplots**: whether or not a household has agricultural plots (1 = yes), and **manure**: how much manure the household used (in kilograms).

Imagine that these are the reasons why data on manure use is missing

- 1002: does not own or operate any agricultural plots
- 1003: did not remember how much manure was used
- 1004: respondent refused to answer the question
- 1006: did not apply any manure, used inorganic fertilizer instead

Table 12.3 *Hypothetical data to illustrate problems with missing data*

housecode	*ownplots*	*manure*
1001	1	10
1002	0	.
1003	1	.
1004	1	.
1005	1	5
1006	1	.

The question of manure use is not applicable for household 1002 and, therefore, this household can be excluded from any analysis on manure use. Clearly 1006 should have 0 for manure use. The other two cases are interesting in the sense that if one were to use quantity of manure used as an explanatory variable, one may wish to do something about 1003 and 1004, for example imputation.[1] Lastly, one may be interested in understanding why respondents refuse to answer certain questions. Without codes allowing for these distinctions, it is not possible to do any of this.

Each program will often have its own codes for missing data; the commonly used ones are 99, 999, or 9999. It is ok to use these codes as long as they do not fall within the range of potential values for the variable, that is, it is not advisable to use 99 to denote missing data for age because there can be people aged 99. For PEN, we used negative numbers as codes for different types of missing data as a solution to this problem.

Observation unit identifier (OUI): It is increasingly common to carry out multi-level surveys. One may carry out village surveys for contextual information, household surveys for the household data and plot level surveys for agricultural production data. Moreover, under each category, multi-surveys may be conducted. As an example, to shorten the recall period, PEN chose to use quarterly surveys to collect income data.

In order to combine data from the different surveys, there needs to be a common variable and code that allows the surveys to 'talk to each other'. This code is the observation unit identifier, a unique code assigned to each household/village and consistent across all surveys in which the village/household appears. It is advisable that this is a number for the very problems related to text data that we discussed above.

Other: It is common to use the code 'other' for all responses that fall outside known ones. However, while the 'other' code is useful to avoid gaps in the data, it is not helpful if a large proportion of your responses are classified as 'other'. Imagine a table, which shows that the most important crop, contributing 50 per cent of the sector income, was 'other'. Codes are cheap. It is better to enter as many unique answers as possible, by assigning new codes. Later, during the analysis stage, it is easy to re-classify or re-code these into fewer aggregate codes or the other code.

Hierarchical coding: A useful way to create codes is to think of them as being part of a hierarchy that starts with a high level category, a parent code, which is passed on to the subsequent subcategories, child codes. As an example, forests could be coded as 100, natural forests as 110 and closed natural forests as 111. These types of hierarchical codes introduce structure that results in better organization of data. In addition, because the child codes contain information about the parent codes, they give users more flexibility on the level of detail for analysis. However, this must be used with some care, as codes risk becoming very long (with higher risks of errors in data entry).

Maintenance of the codebook: It is advisable to have one person responsible for creating, updating and maintaining the codebook. This ensures that there is no duplication of codes. It is definitely not a good idea to have codebooks developed separately with the hope that they will be merged into a single codebook.

Coding on or off the fly: Coding on the fly or coding as you go refers to the situation where one assigns codes concurrently with data entry. As discussed, this is inefficient and error prone. It is advisable that all responses are coded before data entry begins to ensure a smoother and less error-prone exercise.

Data entry

The objective of data entry is to accurately transcribe data from the questionnaire to the computer and it should not be confused with data management or data analysis. This may sound obvious, however, it is not uncommon to see data entry sheets such as the one in Figure 12.2 that, in addition to the raw data, may have aggregated data, a codebook and may even include charts. In Figure 12.2, columns A and B are the raw data, D and E are aggregate data, and F–M is a chart plotting the data in D and E. Doing this is a certain recipe for data quality problems and is highly discouraged.

The conventional way to enter data is in what is referred to as a rectangular format, also called the data matrix, illustrated in Table 12.4, where columns are variables (or fields). These generally correspond with questions on the questionnaire. Rows are observations or cases, which would correspond to the observation units of the study (for example, households in a household survey or

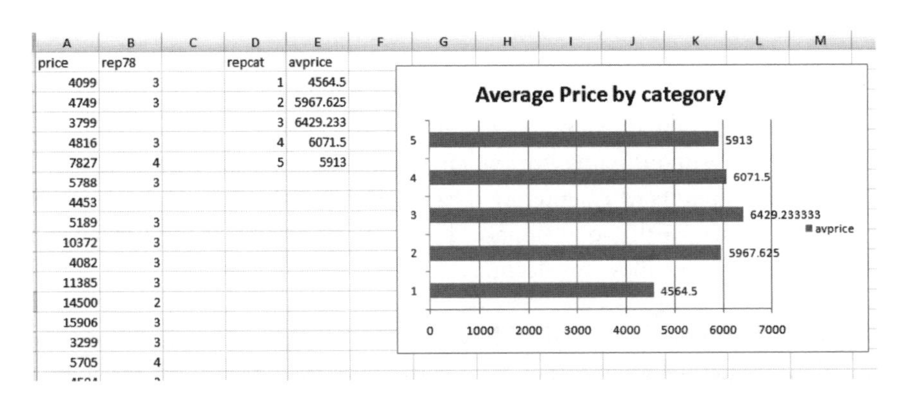

Figure 12.2 *An example of how not to proceed: Mixing raw data, aggregate data and data analysis*

Table 12.4 *Example of a conventional data matrix*

Household	*Age*	*Sex*	*Education*
1001	79	0	1
1002	62	1	1
1003	48	1	4
1004	62	0	0

villages in a village survey). Cells are the data on a given variable for a specific observation unit.

In designing the system for data entry, the following should be kept in mind:

Ease of use: It is pointless to have a 'world-class' data entry system that you cannot use, as this will not only make the data entry process longer, you are also more likely to make data entry errors. That said, what may appear to be an intimidating system may turn out to be easy to use once a user has familiarized themselves with it and so it is worthwhile to try to acquaint yourself with a new data entry system and make a decision based on your experience with it.

Restrictions: A good data entry system should allow a user to set restrictions on what type of data is entered in a given column or cell. Examples of restrictions include:

- **Variable type**: This means text (string) or numeric. Once a variable is defined as numeric, you should not be able to enter text, however, the reverse is possible. This also means that once a variable is defined as numeric, you cannot enter 'NA' for not applicable and will thus need to create a code for it.
- **Range of values**: As discussed earlier, one of the advantages of working with numeric data is that you can restrict the valid entries for a variable to a range or list. As an example, if the variable 'maristat' (marital status) took on four values, married, single, divorced, widowed, then the list of valid entries could be defined to take on values 1, 2, 3 and 4.
- **Precision**: Without going into much detail here, there is no point allowing for decimal points when entering data on the marital status, as all entries are integer values with no decimals. However, for example, the quantities harvested should allow for decimals.

Graphical user interface (GUI): Most data entry programmes will have, as the underlying structure, some sort of spreadsheet that stores data in the rectangular format. A few programmes allow the user to design an interface which can ease data entry. This interface could in principle be designed to look like the questionnaire which then gives familiarity and therefore speeds up the data entry process.

Which software to use?

Word processors (such as Microsoft Word) or text editors (such as Notepad) can be used to record and store data, however, these are more suited to recording free-flowing text. It is possible, in word processors, to get the rectangular format by creating a table, but that is all it will be, a table without any of the important features discussed above. As such, word processors and text editors are not recommended for data entry. Some statistical programs such as SPSS,[2] allow for data entry and may be used, however, the most commonly used platform for data entry is spreadsheets (for example, Microsoft Excel, Lotus, CALC). The main advantages of spreadsheets are that they have the desired rectangular format by default, their cell contents can be restricted,[3] and users can design and add drop-down menus with pre-coded responses. Their biggest advantage is probably their familiarity. Almost all computer users have had some experience with at least one spreadsheet program and therefore find them easy to use, however, this 'advantage' is their Achilles heel because they are as easy to abuse and must therefore be used with great care. Stories abound of data catastrophes resulting from careless use of spreadsheets.[4]

Arguably, the best way to enter data is to use a database. The next logical question is, what is a database and how does it differ from the spreadsheets discussed above? According to Harrington 'there is no term more misunderstood and misused in all of business computing than database' (2009, p4) and yet because their importance, there is no shortage of material on them. A comprehensive treatment of this material is covered in database theory, which is beyond the scope of this chapter.

For this chapter, we use the term database to refer to a mechanism that is used to store information, or data (Stephens and Plew, 2001). As with spreadsheets, databases allow users to enter, modify, delete and retrieve data, however, they have additional advantages, which include restrictions on cell contents, the capability of designing a graphical interface to ease data entry and security of data through restrictions on access. Beyond these, one of the most important attributes is the capability of using multiple but related tables to enter data. This is made possible by what is referred to as a primary key, which is the combination of one or more column values in a table that make a row of data unique within the table (Stephens and Plew, 2001). Multiple tables store data efficiently and effectively (Schulman, 2006). For more on multiple tables and the advantages of databases, see Schulman (2006).

Data checking/cleaning

'I found mothers younger than their children. I found men who claimed to have cervical smears' (Anonymous, 2007). It is hard to overstate the importance of data checking/cleaning and how distressing it can be. Anonymous (2007) writes that real data gives the angst of wondering whether the people who fill in forms bear any resemblance to the great mass of humanity or whether they got the dog to tick the boxes.

It is not that the enumerators and data entry clerks bear no resemblance with the great mass of humanity. Rather, this can be thought about as the classic principal–agent problem where, if you do not monitor quality and link it to pay, quality plays second fiddle to quantity (data entry clerks are usually paid per questionnaire). Data checking is a way of monitoring quality. Visual inspection is the most basic form of data checking but as the human eye can only see so much, the bulk of data checking should be done by computers, which are faster, more accurate and more thorough. Data should be checked for, among other things, duplicates, illegal entries, missing values, inconsistencies, and outliers.

Duplicates: As discussed above, it is important that each observation unit is uniquely identified, particularly if you work with multiple tables. This ensures that data are assigned to the right unit. In relational databases, this is achieved by the primary key which uniquely identifies each record. However, if a relation database is not used, it is important to check for that there are not duplicates in the record identifier.

Illegal entries: Generally, numeric variables are either discrete or continuous. Discrete variables are those that have a finite set of values, for example, the variable sex will generally take on two values, male or female (which can be coded as 0 and 1 respectively). Continuous variables are those that taken on any range of values. Both types will have illegal entries but it is much easier to identify illegal entries in discrete variables because the set of values is known apriori. Illegal entries can be checked using frequency tables, graphs or queries.

Frequency tables

Given that we know all the possible values a variable can take on, for example, the variable hhc_sex (sex of member) with two values, 0 and 1, a simple table of

Table 12.5 *Example of uncovering illegal data entry using a frequency table*

Sex of member	*Frequency*	*Per cent*	*Cum.*
0	579	53.02	53.02
1	512	46.89	99.91
3	1	0.09	100
Total	1092	100	

frequencies will show if there are any illegal entries. In Table 12.5, we see that we have an illegal entry, 3.

Graphs
The same information above can be seen graphically as illustrated in Figure 12.3.

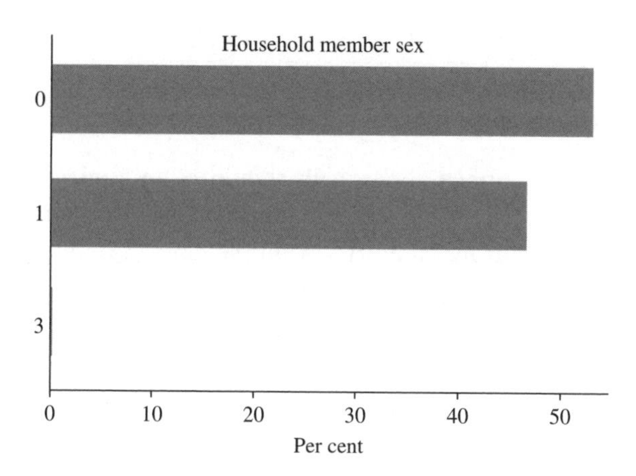

Figure 12.3 *Example of uncovering illegal data entry using graphics*

Queries
While both the table of frequency and the graph can show that you have some illegal entries, they are not efficient for at least two reasons:

i. **Unnecessary information**: When identifying illegal entries, you are not necessarily interested in the distribution of the valid entries and yet both the table of frequencies and the graph show this information.
ii. **Culprit not revealed**: While both tell us that we have an illegal entry, 3, they do not reveal which household/member has this illegal entry so you still have to go an extra step to identify the household.

It is therefore better to query the data and print out the offending observations. Queries take the general form:

1. ACTION: select, list, print
2. VARIABLE: the name of variable(s) you would like to check
3. SOURCE: the source of data, typically a table.
4. CONDITION: what you consider an error, e.g. values outside a list or range of expected values

As an example, if we wanted to see which household, and which member of the household, has the variable sex entered as 3, the Stata command and output to identify the culprit unit would be:

```
list housecode  hhc_pid hhc_sex  if hhc_sex < 0 1  hhc_sex > 1
```

	Housecode	hhc_pid	hhc_sex
583.	317	7	3

Figure 12.4 *Example of Stata query to find illegal data entry*

Missing data: As mentioned above, gaps should be minimized and codes for missing data should be used. Nonetheless, during data entry, some cells may be skipped in error. As with illegal entries, a simple table of frequencies can show missing data. However, a query will identify the specific offending observation.

Inconsistencies: It is common for questionnaires to have interrelated questions, where one question leads to another or where some questions depend on the response given to previous questions. See Table 12.6 for an example from PEN's A1 survey (see Chapter 1).

You should check that all migrant households (who answer no (0) in question 3, Table 12.6) report how many years they have lived in the village.

Outliers: Most data, particularly cross-sectional survey data, contain unusual observations, unusual in the sense that they are inconsistent with the bulk of the data being either too large or too small. Such observations are commonly referred to as outliers.

The most used technique for identifying outliers is the boxplot proposed by Tukey (1977). This method uses 'fences' as criteria to label an observation an outlier with values less or greater than the lower upper fences respectively classified as outliers. The fences are defined as $L_f = P(25) - k*IQR$ and $U_f = P(75) + k*IQR$ for the lower and upper fence respectively, and where $P(25)$ and $P(75)$ are the 25th and 75th percentiles, IQR is the interquartile range $(P(75) - P(25))$ and k is an arbitrary number. In general, a value is a mild outlier if it falls outside (above and below) the fences given by $k= 1.5$ and an extreme outlier when $k = 3$.

However, Hubert and Vandervieren (2008) note that when the data are skewed, usually too many points are classified as outliers. We found this to be a real and big problem in the PEN data set, for example, price data tended have spikes/lumps because of very little variation in product prices for any given product unit combination. Our data were also skewed, for example, data on agricultural input use tended to be skewed to the left with the bulk of the households using none or very small quantities. Because of this, we wrote a Stata program – obsofint – (Buis and Babigumira, 2010), an implementation

Table 12.6 *An example from PEN's A1 survey*

3. Was the household head born in this village? *If 'yes', go to 5.*	*(1-0)*
4. **If 'no':** how long has the household head lived in the village?	*years*

of Tukey's fences combined with the rule of huge error outlier labelling method with some modifications to account for skewed or lumped data.

Conclusions

A good quality data set is a basis for useful information that informs policy and aids decision-making. Practically, a good quality data set lends itself to multiple visits by the same or different users, likely yielding new insights with each visit. In contrast, bad quality data may lead to wrong interventions, is frustrating to use, and may ultimately be rendered unusable and discarded, which would be a highly undesirable outcome given the costs and time spent collecting the data.

While data entry and checking are further downstream in the research process, they should not be dealt with in an ad hoc manner, rather, they should be incorporated at the outset so that they can inform and be informed by the research process. Do not wait until the survey is done to begin thinking about data entry and storage. As you design the questionnaire, think about how data will be entered and stored and, if possible, concurrently design your data entry system.

It is better to enter numerical rather than text data so try to code all categorical responses prior to data entry. One person should maintain the codebook and existing codes should never be overwritten. Instead, add new codes and do not worry about the size of the codebook. Try not to create aggregate codes at the coding stage as you can easily do this when the data have been entered. Remember that aggregating data comes at the cost of loss of detail. You can aggregate codes but you cannot disaggregate them once the data have been entered.

In trying to get good quality data, the winning strategy is 'prepare and prevent'. Think about the potential sources of error and try to put in place mechanisms to minimize the errors. Lastly, leave a paper trail. Documentation improves data quality so record as much as you can while you still remember.

Key messages

- Use codes instead of text, if possible.
- Data entry is about entering data and should not be mixed with data analysis.
- Ensure that all observation units are assigned unique codes.
- Don't wait until the end of all field work to start thinking about data entry.

Notes

1 See www.multiple-imputation.com for some references.
2 www.spss.com/software/data-collection/data-entry.
3 See, for example, http://spreadsheets.about.com/od/datamanagementinexcel/ss/excel_database.htm.
4 See, for example, www.burns-stat.com/pages/Tutor/spreadsheet_addiction.html.

References

Anonymous (2007) 'Dr Fisher's casebook: The trouble with data', *Significance*, vol 4, no 2, p74

Audit Commission (2007) 'Improving information to support decision-making: A framework to support improvement in data quality in the public sector', Audit Commission for local authorities and the National Health Service in England, London

Buis, Maarten, L. and Babigumira, R. (2010) 'OBSOFINT: Stata module to display observations of interest', Statistical Software Components S457175, Boston College Department of Economics, http://ideas.repec.org/c/boc/bocode/s457175.html (version 1.0.4), accessed 20 December 2010

Gladwell, M. (2008) *Outliers: The Story of Success*, Allen Lane, Camberwell

Harrington, J. L. (2009) *Relational Database Design and Implementation: Clearly Explained*, Morgan Kaufmann, Burlington, VT

Hubert, M. and Vandervieren, E. (2008) 'An adjusted boxplot for skewed distributions', *Computational Statistics and Data Analysis*, vol 52, no 12, pp5186–5201

Lee, Y. W., Pipino, L. L., Funk, J. D. and Wang, R. Y. (2006) *Journey to Data Quality*, MIT Press, Cambridge, MA

Redman, T. C. (2001) *Data Quality: The Field Guide*, Butterworth-Heinemann, Oxford and Woburn, MA

Refaat, M. (2007) *Data Preparation for Data Mining Using SAS*, Morgan Kaufmann, San Francisco, CA

Schulman, J. (2006) *Managing Your Patients' Data in the Neonatal and Pediatric ICU: An Introduction to Databases and Statistical Analysis*, Blackwell, Oxford

StataCorp (2009) *Stata Statistical Software: Release 11*, StataCorp LP, College Station, TX

Stephens, R. K. and Plew, R. R. (2001) *Database Design*, Sams Publishing, Indianapolis, IN

Swift, B. (2006) 'Preparing numerical data', in Sapsford, R. and Jupp, V. (eds) *Data Collection and Analysis*, Sage, London

Tukey, J.W. (1977) *Exploratory Data Analysis*, Addison-Wesley, Reading, MA

Chapter 13

An Introduction to Data Analysis

Gerald Shively and Marty Luckert

Since the behaviors to be sought are unknown, computers cannot be instructed to watch out for them. Computers can 'keep track' of a complex of behaviors, but only the human mind can discern. R. Buckminster Fuller (*Intuition*, 1972, Doubleday, New York)

Introduction

Data analysis is generally undertaken to tell a story. The researcher is seeking answers to questions that when revealed, will captivate the audience in terms of their insights and relevance. The process of analysing the data, in a way that tells a compelling story, requires the researcher to pursue two general aims: to describe and to explain. The first aim of data analysis, and the one that constitutes the initial stage of analysis, is to accurately describe a situation of concern. Descriptions may be used to characterize patterns of behaviour and outcomes observed among respondents in the research sample. For example, the goal might be to better understand who uses and relies on forests – whether it is the rich or the poor, the young or the old, long-term residents or recent migrants. Provided one has collected the right information, one should be well-positioned to answer such questions.

Describing what has been observed is a laudable and necessary goal, since without a basic understanding of the data one may not be confident in her ability as a researcher to cast light on the issues that motivated the research in the first place. Furthermore, describing a situation that is not well-studied is in itself worthwhile. Descriptive analysis allows one to answer questions regarding the *who*, the *what*, the *where*, the *when* and the *how*.

For most forms of social science research, however, description alone will be insufficient for telling a compelling story. The underlying question of *why* one observes certain patterns in the data will be the one that pushes the research agenda forward. This second and deeper motivation for undertaking data analysis is to explain reasons behind patterns of behaviour and outcomes observed among respondents in the research sample. In fact, many researchers regard these types of explanation as the highest calling for a researcher, especially if one is interested in conducting research that is relevant to policy-makers. So, for example, if one finds that the poor rely on forests more than the rich, or the young rely on them more than the old, the background causes are important. Do the observed patterns reflect low agricultural capacity, surplus labour in the household or something else? If the goal is to communicate research findings in a way that maximizes the opportunity to have influence and impact, *explanation* is the litmus test for research.

Though data analysis can take many forms (see Box 13.1), our emphasis in this chapter is on statistical methods. To the extent the sample is representative of some larger population (see Chapter 4), statistical analysis can allow one to use the knowledge of the sample to generalize about what is happening in the underlying population of interest. Because there are many statistical approaches, of varying degrees of sophistication, it is neither feasible nor appropriate for us to provide in this chapter a set of comprehensive guidelines for conducting data analysis. Instead, we introduce some ideas to get started and illustrate some basic approaches using examples from data sets that we have used in our own research.

We begin by discussing ways to describe and explore the data, including the use of statistical tools and graphical approaches. We then discuss how hypothesis-driven analysis can be used to lead descriptive analysis towards explanatory analysis. Unfortunately, the use of statistics for explanatory analysis presents numerous common pitfalls. We briefly discuss some of these and then conclude with some final tips.

Box 13.1 *What type of analysis?*

The assumption maintained in this chapter is that the reader is interested in *statistical analysis*, which involves analysis of population characteristics by inference from sampling. This approach typically involves the calculation of sample statistics and the application of statistical tests to establish the validity of hypotheses or the strength of hypothesized relationships.

Other approaches to analysis that seek to explain observed patterns in data include the use of simulations, optimization methods, case studies (that allow analytical, rather than statistical generalization) and more qualitative approaches.

Exploring and describing the data

Exploratory analysis typically begins with the calculation of basic descriptive statistics. This may rely on univariate methods (studying a single variable observed in isolation), bivariate methods (examining the relationship between a pair of variables) or multivariate methods (studying associations among a group of variables). For continuous variables, descriptive univariate statistics include measures of central tendency (such as means, medians and modes), as well as measures of spread (such as ranges, variances, standard deviations and coefficients of variation). These indicators provide insights into the data and give a picture of 'typical' or 'average' characteristics or performance, as well as a rough idea of dispersion in the data. In addition to computing such sample statistics, simple graphical methods can be used to explore the way values of a variable are distributed within a sample. Frequency and percentage histograms (or kernel density estimators) are used for this purpose and provide a very useful way to visualize the data, either to understand how data may be clustered or grouped, or to identify observations that take on extreme values, and which are therefore worthy of further investigation. As an example, Figure 13.1 illustrates the use of a histogram to display the distribution of forest clearing among a sample of 121 agricultural households in the Philippine province of Palawan.[1]

The histogram in Figure 13.1 helps us to visualize that the majority of households in the sample (nearly half) cleared little or no land, and that the remainder cleared a modest amount ranging up to 2ha. The histogram provides information that would be hard to ascertain from looking at the mean (0.38ha)

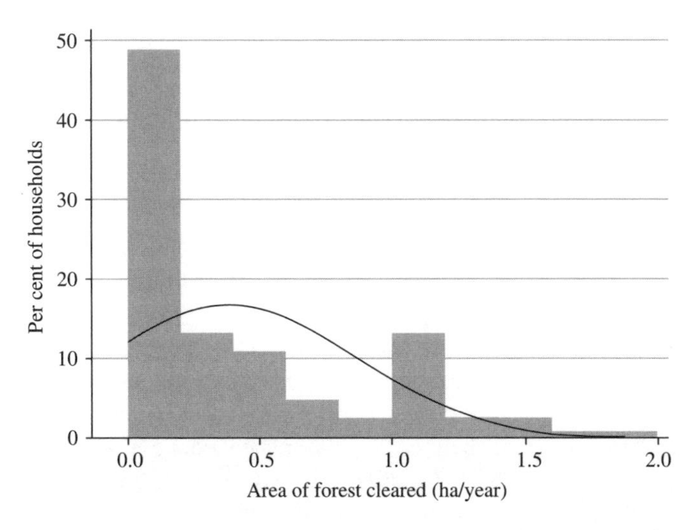

Figure 13.1 *Forest clearing among agricultural households in the Philippines, 1995*

and standard deviation (0.48ha) alone. The curved line superimposed on the histogram is an estimate of what the distribution would have looked like had it been distributed following a normal, or bell-shaped, curve. Among the things we learn from Figure 13.1 are that the variable is not normally distributed (because much of the mass of the distribution is located at zero), and that the response '1ha' appears more frequently than one might expect (quite possibly because of an anchoring effect in which households tended to report 1ha as an approximation to an amount close to but not exactly 1ha).

At the initial exploratory stage of analysis, it is useful to generate simple descriptive statistics and frequency histograms for all continuous variables in the data set. This is a good way to begin to understand the characteristics of the data. For categorical variables – those that take on discrete values – tables of frequencies or proportions can be used to explore the data. This kind of basic exploratory analysis is often combined with data cleaning in an iterative manner, in which the results from descriptive or visual analysis are used to investigate observations of particular interest or concern in the data set (see Chapter 12).

Initially, the exploration of the data will rely on unconditional measures. Statistics are said to be 'unconditional' unless the researcher has partitioned (conditioned) the data along some particular dimension represented by some other observed factor in the data set. As an example, one might think about farm size in an overall sample of farm households (in other words, the unconditional mean and standard deviation) and contrast this with farm size for female-headed and male-headed households (the mean and standard deviation conditional on the sex of the household head). Table 13.1 contains exactly these descriptive data for a sample of 380 households in rural Malawi.[2] The data illustrate that farm size among male-headed households is somewhat higher than among female-headed households (1.7ha versus 1.1ha). The coefficient of variation, a measure of spread defined as the standard deviation divided by the mean, suggests that farm size is also somewhat more variable among males. Whether

Box 13.2 *Choosing software*

When undertaking data analysis, researchers often have a wide range of statistical software packages from which to choose. Most packages support many of the same types of statistical tests. Therefore the choice of what software to use may depend on other considerations, such as availability, familiarity, cost and access to support. But for some applications (for example, time series analysis or the use of limited dependent variables), more specialized software, or software that runs specific sub-programs that other researchers have developed, may be useful.

Table 13.1 *Farm size (in hectares) among a sample of rural households in Malawi*

	Mean	Standard deviation	Coefficient of variation	Number of observations
All households	1.6	1.5	90%	380
Female-headed	1.1	0.6	59%	56
Male-headed	1.7	1.5	88%	324

higher variance among male-headed households represents more observations above the mean, more observations below the mean or more of both, cannot be determined from the data in Table 13.1. Instead, one might need to go back to the data and plot separate histograms for male-headed and female-headed households in the data set.

Importantly, the forgoing example illustrates that univariate measures alone may mask important differences in one's data. For this reason, it is frequently useful to graduate to the use of bivariate statistics as a way of looking at two variables in combination. For continuous variables, choices include bivariate scatter plots between variables that are believed to be related, or the computation of pair-wise correlation coefficients among variables. Figure 13.2 is a scatterplot showing observed combinations of labour and maize output for upland maize farms in the Philippines (part of the same sample used in Figure 13.1).[3] The scatter plot uses a dot to represent each labour – output combination in the data set, and therefore shows the empirical bivariate relationship between input (labour) and output (maize). A simple linear regression line has been superimposed over the cloud of data. This kind of plot is useful for looking at

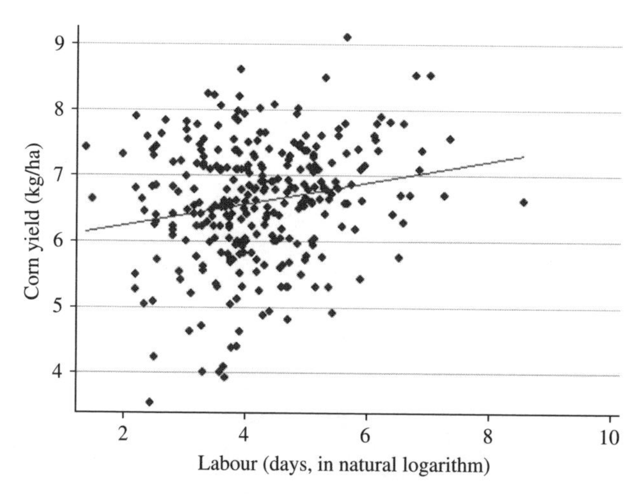

Figure 13.2 *Labour and output among agricultural households in the Philippines, 1995*

the potential directions and magnitudes of association in the data. It can also be helpful for identifying questionable observations that might not look unusual in the context of a histogram but might be unusual in a bivariate setting. For example, high yields might not be surprising in isolation, but might be a concern if they are associated with low levels of labour or fertilizer use. As above, such observations might be candidates for further investigation.

In some cases, more powerful graphing features can be used to examine statistical relationships between two continuous variables in greater depth. For example, Stata provides a number of ways to produce graphs that include statistical confidence intervals around a fitted regression line. Figure 13.3 is a polynomial regression based on data from two districts of Western Uganda. The figure, which is based on 284 observations, shows the relationship between farm size and the share of household income derived from forests.[4] The figure is suggestive of a U-shaped relationship in which households with the smallest and largest farms derive larger shares of income from charcoal production than those with landholdings in the middle of the distribution.

Before concluding too much from Figure 13.3, however, we should consider the issue of statistical confidence. Statistical confidence simply refers to how certain we can be that a particular outcome or pattern in the data arises by 'more than chance'. Statisticians usually refer to confidence in terms of confidence intervals (CIs). The shaded area around the line in Figure 13.3 is a 95 per cent confidence interval for the relationship indicated by the line. The shaded area is rather narrow at the left-hand side (corresponding to observations of small farms) and rather wide at the right-hand side (corresponding to observations of large farms). One benefit of including the 95 per cent confidence interval as part

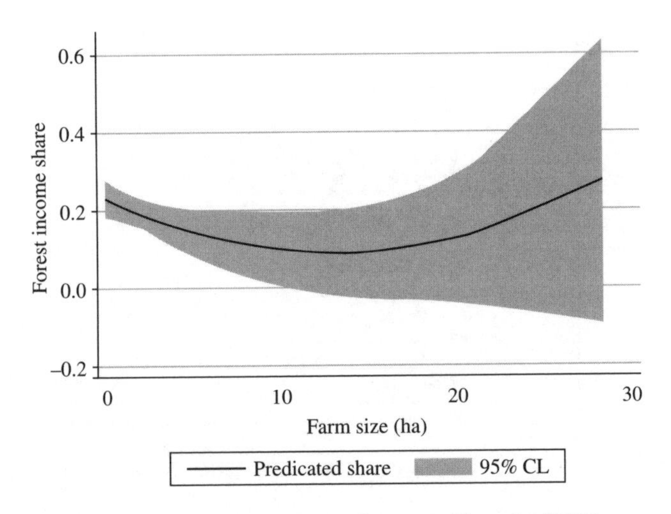

Figure 13.3 *Farm size and forest income shares in Uganda, 2008*

of the plot is that we can immediately see that the purported U-shape relationship is statistically weak at the upper end of the farm size distribution. The reason for the wide confidence interval, it turns out, is that relatively few observations with large farm sizes are seen in the data. As a result, the bivariate relationship in this region of the data cannot be estimated with much statistical precision and one cannot really rule out the possibility that the relationship is different (and perhaps even monotonically decreasing). Although the data are attempting to tell a story, in this case, we must be careful about reading 'too much' into the data. Another way of saying statistical confidence is low, is to say the data provide limited support for the story. As always, however, absence of evidence is not necessarily evidence of absence. As was indicated in Figure 13.3, there may be cases where there are too few observations, creating wide confidence bounds, which confound the ability to conclude whether a relationship exists. Moreover, true bivariate relationships sometimes can be masked by confounding variables that have not been accounted for.

Hypothesis-driven analysis

Establishing clear and testable hypotheses can be the most important part of the research, as argued in Chapter 3. Establishing hypotheses is an integral part of the story that one will be telling. A good hypothesis keeps the analysis and the audience's attention focused on policy-relevant implications that arise from the data. Moreover, a well-specified hypothesis creates valuable information after the results of the statistical tests are known.

If this 'hypothesis-driven' stage of analysis is outlined well, then one will know what one is doing, how one is going to do it and why. An example of a clear and testable descriptive hypothesis is the following:

> *Low-income households rely more heavily on forest-based income, in relation to their total income, than households with higher incomes.*

This hypothesis is said to be 'testable' or 'falsifiable' because data can be used to support either this conjecture or the alternative, namely that low-income households are no more or less reliant on forest-based incomes than those with higher incomes.

While stating our research question in the form of a testable hypothesis is useful, it does not necessarily provide an unambiguous way forward. To sharpen this hypothesis further, it would probably be necessary and useful to clearly identify some variables and concepts. For example, what do we mean by 'rely on'? How are we defining income? What constitutes forest-based income? What do we mean by 'low' and 'high' incomes? Only after we have carefully defined the variables that we are using, and clearly articulated the underlying concepts, can we be sure about what we are actually testing. Defining variables and concepts clearly and carefully, and communicating those definitions and concepts to readers is a constituent part of data analysis. The importance of accurately constructing and defining variables cannot be overemphasized, since a hypothesis test that relies on the use of flawed or misunderstood variables can be misleading and therefore worse than no test at all.

From a broader perspective, if the ideas regarding the hypotheses are not clear, or the underlying concepts are muddled, then one will likely flounder with little basis for directing the inquiries. For example, a less clear statement of the hypothesis provided above could be something such as 'forests are important to households'. In the first instance, one could calculate proportions of different types of income as part of household portfolios, and test whether they are significantly different for sub-groups of the sample (for example, low-income and high-income households). In the second instance, the hypothesis is not stated in a clear enough way. For example, it is not clear how 'important' is defined, so it is not clear how the importance will be measured or related to forests. The second statement is actually more akin to a general research question (for example, 'are forests important to households?') than a falsifiable hypothesis.

Even in the more sharply constructed form, however, our example of a clear and testable hypothesis, remains rather general. Though such a descriptive hypothesis can be useful, as mentioned above, scientists are generally not satisfied just describing what is. Rather, they are usually more interested in understanding why things are as they are. Along these lines, a clear explanatory hypothesis would be:

> *A shortage of agricultural land among poorer households leads them to rely more heavily on forest-based incomes compared with relatively richer households.*

An explanatory hypothesis such as this typically postulates cause-and-effect relationships among multiple variables. In this multivariate setting, at least three variables would be necessary to support a test of the hypothesis: agricultural land holdings, forest-based income and a measure of wealth and/or income. More realistically, a number of additional variables (for example, location, household size, education) would also be included in the analysis to control for conditioning factors that could also influence the relationships among the variables of interest. Multiple regression analysis is the approach most frequently relied upon for this kind of analysis.

Using regressions to analyse cause-and-effect relationships among multiple variables involves making choices regarding what variables to include and evaluate in a regression. On what basis should these choices be made? In most social sciences, cause-and-effect relationships arise from theory. In economics, theoretical models can provide a rigorous basis for the types of variables to include in analyses. For example, theory related to supply and demand suggests that both of these are dependent on prices and quantities. Formal theoretical models are used to specify hypothesized cause-and-effect relationships, made up of dependent and independent variables. These more simplified relationships that arise from complex theoretical models are frequently referred to as 'reduced form' equations. For most empirical studies in economics, theory has already been developed and reduced forms have already been estimated for many types of empirical relationships. Therefore, to establish a hypothesis, one can sometimes rely on theoretical and empirical work that has already been done in other places.

To sum up, hypothesis-driven data analysis is often the preferred way of pursuing research questions, but must also be practised cautiously as it can constrain and/or bias data analysis. A researcher needs to keep an open mind regarding alternative approaches and the interpretation of unexpected results. In the next section, we review a number of key concepts that can help one to sift through the complex relationships in the data that are formed into the story.

Common statistical pitfalls in interpreting the story in the data

Interpreting data largely arises from trying to identify patterns that may or may not support the hypothesis. But an accurate interpretation of those patterns may be obfuscated by a number of different issues. Though each of these issues represents a large area of inquiry in econometrics, below we talk about some things to look out for and briefly introduce some possible solutions.

One key issue in analysing data is the distinction between correlation and causation. Correlation implies that two variables (after controlling for other effects in a multivariate setting) are related, or move together, either positively or negatively. Given the universe of variables that can be measured, we would expect to find correlations between many variables. But a key consideration is whether these relationships are actually measuring relationships that are meaningful (Box 13.3). That is, are the correlated variables supporting an explanatory hypothesis, or do they just happen to be moving together by chance? Continuing with the example from above, we may find that households

Box 13.3 *Measuring impacts*

The desire to move beyond correlation to assess causation raises some tricky issues related to statistical design. The gold standard for measuring impacts is the randomized controlled experiment in which experimental subjects are randomly assigned to treatment and control groups, and variables of interest are measured before and after the treatment in order to assess treatment impact.

Unfortunately, it is rarely the case that researchers can conduct randomized controlled experiments on variables of policy interest or concern. Instead, a variety of methods have been developed in an attempt to measure programme or policy impacts in the absence of randomization. The main approaches include regression on covariates, use of instrumental variables techniques, matching methods and regression discontinuity designs.

These methods are described in greater detail by Angrist and Pishchke (2009).

with smaller amounts of agricultural land, and those with more household members born in even-numbered years, rely more on forest-based incomes than their sample cohorts. Though both of these variables may be correlated with a heavier reliance on forest-based income, the second correlation is likely 'spurious'. Only the first could be reliably hypothesized to influence our variable of interest in a causal manner. Further to our discussion on hypothesis-driven research above, the variable on access to agricultural land has a theoretical foundation that establishes the hypothesis, while the variable on even-numbered years does not. The key point is that correlation is a necessary but not a sufficient condition for drawing helpful conclusions based on one's analysis. We are ultimately interested in causation, which combines correlation with a logical basis for hypothesizing cause and effect, rather than correlation alone.

Closely related to the issue of causality is the distinction between dependent and independent variables. Strictly speaking, dependent variables belong on the left-hand side of a regression and independent variables (those determined independently of the outcome variable) belong on the right-hand side. Unfortunately, the term 'independent' is often used loosely (and mistakenly, in our view) to refer to right-hand side variables in a regression, regardless of whether they are truly independent of the outcome variable of interest. Sets of independent variables are sometimes referred to as control variables, because they are understood to be factors that one controls for when exploring other bivariate or multivariate relationships. For example, even if we are only interested in the bivariate relationship of whether a donor project has increased agricultural income, we would nonetheless likely need control variables that correct for differences in other factors that could have influenced income, such as weather. Control variables may also be thought of within the context of conditional means. In the example above, the weather control variable is simply the variable that we condition on when calculating a conditional mean of agricultural income.

If one has not previously encountered the distinction between endogenous and exogenous variables, then it is important to think about this concept as well. Exogenous variables are those that are assumed to be outside the context that is being studied, and therefore assumed to not be affected by the behaviour underlying the explanations that emerge in the analysis. For example, fluctuating weather or changes in national policies may influence agricultural income, but they are generally assumed to be exogenous because they are not thought to be influenced by the individuals in the study. Note, however, that the scale of the analysis could influence the credibility of the assumption regarding whether a variable is exogenous. For example, if one is considering the agricultural production behaviour of people within an entire country, then the aggregated results of this large group could influence national policies, creating doubts regarding whether changes in national policies are exogenous.

Exogenous variables share a lot of features with independent variables. In contrast, endogenous variables are those that are typically co-determined with other variables examined in the analysis.[5] Strictly speaking, endogenous variables belong on the left-hand side of a regression, with exogenous variables (their causal forces) appearing on the right-hand side. However, in some settings it may be the case that one wishes to place a variable on the right-hand side of a regression that is endogenously determined along with the dependent variable. In this case, econometricians have developed a number of approaches to deal with this situation. A short example is useful for illustrating the thought process that underlies how this issue might be addressed.

Consider the goal of measuring whether a household's use of forest resources influences decisions regarding the use of agricultural inputs (such as improved seed or fertilizer). One hypothesis might be that income from the sale of forest products facilitates the purchase of agricultural inputs, since cash from the former may be needed to purchase the latter. An alternative hypothesis could be that households tend to specialize in their activities, and therefore that use of forests tends to compete with agricultural activities. In this case, forest income will be negatively correlated with input purchases. The central problem with investigating this issue is that decisions to use forests and decisions to purchase agricultural inputs may be made jointly or contemporaneously, in a way that each influences the other within a specific window of time. This might be because both decisions have implications for the use of household labour, and the allocation of labour to one activity is related to the allocation of labour to another. In other words, one cannot treat either factor as exogenous to the other; we say they are 'endogenously determined'. Strictly speaking, the inclusion of one variable, say forest income, as an explanatory variable for the other, input purchases, violates an important assumption upon which Ordinary Least Squares regression (OLS) is predicated.[6]

One way to approach the analysis of this question might be to measure the variables of interest at different points in time. If forest income is observed in period 1 and input purchases are observed in period 2, then forest income might be reasonably considered exogenous from the point of view of period 2, and one could estimate a regression with period 2 input purchases as the dependent variable and period 1 forest income as the independent variable.[7]

Another approach would be to search for a variable to serve as an 'instrument' for forest income. Such a variable is called an instrumental variable if it is (a) highly correlated with forest income, but (b) uncorrelated with the error term associated with the input purchases. What kind of variable might serve this purpose? One candidate might be proximity to the forest. If it turns out that distance to the forest is a good predictor of forest income but not correlated with agricultural input purchases, then one could start with a regression in which forest income is explained by distance (presumably higher for those closer to the forest). After running that regression, the predicted values of forest income are then constructed and saved for inclusion in the input purchase regression. This 'instrumented' version can stand in place of the original variable since it contains only the component of income that is related to distance, which we have determined is uncorrelated with input purchases.

Having carefully contemplated issues related to causality and endogeneity in the data set, four additional concerns are worth some examination. First, it is important to define variables precisely and to understand what they represent. We highlighted this idea earlier, but it is always important to clearly distinguish such things as direct measures versus proxies; stocks versus flows and aggregate versus per unit comparisons. Additionally, we emphasize that units of measurement always should be defined. For example, the educational level of a household can be expressed by years of formal schooling of the household head, the average number of years of schooling of all household members or the maximum number of years of schooling in the household. Clearly indicating the units of measure helps others interpret your results and compare/contrast them with other findings in the literature.

Second, some relationships are non-linear. For this reason, it may be useful to consider using quadratic terms for an explanatory variable in a regression (for example, farm size and the square of farm size, under the assumption that farm size might have a diminishing effect on the dependent variable). For the same reason, researchers often use logarithmic transformations of dependent or independent variables in their regressions.[8] Because linear regressions assume a linear relationship between dependent and independent variables, it may fail to uncover a relationship that is non-linear. U-shaped relationships (see Figure 13.3) or relationships with other patterns of strong curvature may be better discovered and studied using transformations of the data.

Third, one should be prepared to check the robustness of the results to alternative specifications or alternative interpretations/hypotheses. In the most general terms, *robustness* refers to how well the interpretation of the data holds up to potential alternative specifications. So, for example, one might be interested in whether an observed regression result is robust to the inclusion of additional explanatory variables, or whether the patterns observed are similar for important subsets of the data. If a wide range of specifications all suggest the same relationship, then one can be more confident about concluding the relationship exists.

Finally, it is important to make sure that the choice of a regression model is properly matched to the form of the dependent variable under examination. Most outcome variables of interest are continuous, and therefore properly suited to OLS regressions. But in other cases, variables may be categorical or take on values that mean OLS is inappropriate. In many instances, in fact, the researcher will have some choice in determining what form of a variable to use for analysis. Consider, for example, a variable to indicate income from forest sources. Table 13.2 presents five possible variables for forest income in a sample of 219 households from Uganda. The first three variables are continuous and, under most situations, could probably appear as the dependent variable in an OLS regression. However, the fact that some households report zero values for cash, subsistence or total forest

Table 13.2 *Measures of forest income among 219 households in Uganda, 2007*

Variable	*Description*	*Mean*
Cash forest income	Total annual household income from sales of forest products (in UgSh)	50,034
Subsistence forest income	Total imputed annual value of forest products used by household (in UgSh)	72,095
Total forest income	Combination of cash and subsistence income (in UgSh)	122,129
Share of forest income	Total forest income divided by total income (share)	0.15
Forest income	Binary indicator of forest income: 0 if forest income $= 0$ 1 if forest income > 0	0.95

income means that these variables are truncated at the lower end of their distribution by values of zero. For this reason, a researcher might choose to use a one-tailed Tobit regression to analyse them. A Tobit procedure fits a non-linear regression line through the data and, in the case of a lower bound, requires all predicted values to fall above the lower cut-off point (here, 0). In the case of the variable for the forest income share, the observed values are, by construction, bounded between 0 and 1. Values of 0 correspond to households with no forest income. Values of 1 correspond to households that receive all income from forests. For this variable, a Tobit model might again be called for, in this case a so-called 'two-tailed' Tobit, with a lower limit of 0 and an upper limit of 1. In general, the more observations in a sample that occur at one or both censoring points, the larger the bias associated with using OLS for the regression. Finally, the last variable listed in Table 13.2 is a binary indicator of forest income. This variable is again bounded between 0 and 1. But in this case the variable can only take on values of 0 and 1. Values of 0 correspond to households with no forest income, and values of 1 correspond to households that reported non-zero forest income. For this variable, the appropriate regression model would be either a Logit or Probit model, both of which predict a household's probability of being either 0 or 1. Which definition of 'forest income' one chooses to use will depend a bit on the context of the analysis. In some instances, a researcher may want to work with models for all forms of the variable, since each tells a somewhat different story about forest income in the sample. The important thing to keep in mind is that the specific form of the variable used may call for a different type of regression model. Fortunately, most statistical packages make it easy to use the right model.

Conclusions and final tips

As suggested above, perhaps the most important thing one must aim to do when conducting the analysis is to listen to the data; that is, allow the data set to speak for itself. This can be especially difficult when one is out to prove or disprove an important point and, for this reason, adopting a high degree of objectivity is essential. The foundation for analysis is a solid and dispassionate understanding of the quality of the data and the context in which it was collected. This understanding will help us to know what can and – of equal importance – cannot be said based on the data. In pursuing the data analysis, one will want to repeatedly return to the storyline, building an evidence-based approach to testing conjectures. As part of this process, it helps to play to the strengths of the data set, while trying to remain objective, honest and self-critical. The data set may present many different and conflicting patterns, and it is important to

Box 13.4 *Passing the 'laugh test'*

Researchers often focus on developing elaborate and complex theories about patterns in their data and then develop hypothesis tests aimed at supporting that theory. But many observers suggest a key way to scrutinize one's analysis is to ask whether it passes the 'laugh test'.

Does the explanation for observed patterns in the data stand up to clear and simple logic? Do informed colleagues find it believable? Would the respondents from the survey find the explanation logical?

If the answer to any of these questions is 'no' or if the explanation for observed patterns is so convoluted that it seems hard to believe (and explain) – in other words, if the explanation might lead people to laugh at its absurdity, then one may need to rethink it, even if the data seem to support the conjecture.

That is not to say that conventional wisdom is always right, or that complicated explanations are wrong. Just keep in mind that clear and simple explanations are often the best.

exercise caution so that preconceived notions do not constrain the analysis. Instead, it is important to think broadly and try to let the data tell the story.

Despite the potential complexity in the data analysis, there are some simple things one can do to increase the rigour and integrity of the analysis. As one struggles with potentially complex explanations, do not overlook more simple ones. Are the stories that are emerging actually believable? The researcher may have gotten so close to the data that he or she has failed to see that what is proposed is incredulous (see Box 13.4).

One way to add perspective to the analysis is to run the ideas past colleagues. Multiple points of view can be crucial in figuring out how to allow the data to tell the story. It can be especially important at initial stages to present the work in lunchtime seminars or small conferences. Young researchers are often afraid to present their work, especially when conclusions are tentative and analysis is ongoing, but these are the conditions that add the greatest value to feedback from colleagues because it is received at a stage where it can be fruitfully used. Whether one presents the work in public or not, it can also be very important to find a reliable colleague who is willing to serve as a sounding board. If both are at early stages in the careers, this can be the start of a mutually beneficial relationship to exchange papers, ideas and constructive feedback.

Another issue that is often overlooked is the importance of keeping good notes during the analysis stage, including a daily log of the activities and well-documented notes on programs and data sets. Although documenting every step of the analysis and every statistical procedure can seem tedious and time-consuming, in the long run such steps can save many hours of frustration if it becomes necessary to retrace the steps or revise and recreate a data file. In addition, the potential for making mistakes when 'working on the fly' is always great. Many researchers can tell horror stories about simple manipulations of their data sets that were undertaken when it was late or they were tired and frustrated. The strategy looked good at the time but proved problematic upon later (and closer) inspection. Being able to easily undo something may require meticulous records. Furthermore, what seems obvious at the time when done may seem much less obvious six months, a year or two years later when a journal reviewer has asked for clarification or an alternative specification of a model. Having good notes and a set of programs that are well-documented may be the deciding factor between getting an article published and having several years' worth of efforts end in frustration.

Key messages

- Make sure your analysis is motivated by a story. This helps to isolate the variables of interest and to construct your analysis in a way that moves from description to explanation.
- Clearly distinguish between correlation and causation. Do not infer the latter from the former and take care to correctly account for any endogeneity that may exist between dependent and independent variables.
- Do not underestimate the power of descriptive analysis, especially graphical analysis. Important relationships may be easiest to discern visually and all econometric work should be preceded by careful visual exploration of the data.

Box 13.5 *Further reading*

There are dozens of good econometric reference books, but three widely used are Wooldridge (2001), Cameron and Trivedi (2005) and Greene (2008). More basic books for beginners include Kennedy (2008), Cameron and Trivedi (2009), Gujarati and Porter (2009) or Wooldridge (2009).

For software, our obvious bias rests with Stata as a statistical package, based on its ease of use, comprehensive documentation and large community of users. Numerous books on how to get started with Stata are available. We think a good place to begin is with *A Gentle Introduction to Stata* by Acock (2006). Baum (2006) and Cameron and Trivedi (2009) also provide very complete overviews of data analysis in the context of Stata, with straightforward nuts-and-bolts foci on regression techniques.

For those seeking inspiration for policy-oriented data analysis, we highly recommend Deaton's (1997) book *The Analysis of Household Surveys: A Microeconomic Approach to Development Policy*, which is lucidly written and squarely focuses on the strengths and pitfalls associated with working with survey data. Finally, we are strong proponents of the use of figures, both in data exploration and in presentation. A fascinating and useful desk reference on this subject is Tufte's (2001) classic.

Notes

1 A number of different statistical software packages are available (see Box 13.2). Figure 13.1 was generated in Stata using the command: **hist cleared_area, percent bin(10) normal**

2 Table 13.1 was generated in Stata using the command: **tab fhoh, sum(farmsize)**

3 Figure 13.2 was generated in Stata using the command: **twoway (scatter lny lnl) (lfit lny lnl)**

4 Figure 13.3 was generated in Stata using the command: **twoway (fpfitci share farmsize)**

5 In the language of econometrics, a variable on the right-hand side of a regression is said to be endogenous if the variable itself is correlated with the error term in the regression. In such a case, an Ordinary Least Squares (OLS) regression – see note 6 – is flawed.

6 OLS regression is a means of fitting a line to data based on minimizing the square of the distance between data points and predicted points on the estimated line.

7 Still, one has to be careful proceeding in this manner. From one perspective, forest income could be considered a 'lagged endogenous variable'. In other words, while the variable itself is observed in the past, the factors that influence it (habit, custom or circumstance) may persist into the future and be just as endogenously connected to the period 2 decision as if the period 1 variable were observed in period 2. Treatment of lagged endogenous variables goes beyond the scope of this chapter. Readers interested in the subject of endogeneity should consult a reliable econometrics textbook, such as Greene (2008).

8 If logarithmic transformations are used, the interpretation of variables is different. The parameter becomes an elasticity that represents a percentage

change in the left-hand side variable for a percentage change in a given right-hand side variable. To ease in interpretation of these models, it is common to take the log transformation of all variables in the model, creating what is termed a log–log or log–linear model.

References

Acock, A. C. (2006) *A Gentle Introduction to Stata*, Stata Press, College Station, TX

Angrist, J. B. and Pishchke, J. S. (2009) *Mostly Harmless Econometrics*, Princeton University Press, Princeton, NJ

Baum, C. F. (2006) *An Introduction to Modern Econometrics Using Stata*, Stata Press, College Station, TX

Cameron, C. A. and Trivedi, P. K. (2005) *Microeconometrics: Methods and Applications*, Cambridge University Press, New York

Cameron, C. A. and Trivedi, P. K. (2009) *Microeconometrics Using Stata*, Stata Press, College Station, TX

Deaton, A. (1997) *The Analysis of Household Surveys: A Microeconomic Approach to Development Policy*, The Johns Hopkins University Press, Baltimore, MD

Greene, W. H. (2008) *Econometric Analysis*, sixth edition, Pearson Prentice-Hall, Upper Saddle River, NJ

Gujarati, D. N. and Porter, D. C. (2009) *Basic Econometrics*, fifth edition, McGraw Hill, New York

Kennedy, P. (2008) *A Guide to Econometrics*, MIT Press, Cambridge, MA

Tufte, E. R. (2001) *The Visual Display of Quantitative Information*, Graphic Press LLC, Cheshire

Wooldridge, J. M. (2001) *Econometric Analysis of Cross Section and Panel Data*, MIT Press, Cambridge, MA

Wooldridge, J. M. (2009) *Introductory Econometrics: A Modern Approach*, fourth edition, Cengage Learning, Florence, KY

Chapter 14

Communicating Research for Influence and Impact

Brian Belcher, Ronnie Babigumira and Theresa Bell

Good communication is as stimulating as black coffee and just as hard to sleep after.
Anne Morrow Lindbergh (1955, *Gift From the Sea*, Pantheon Books)

Communicating research

In the myth of the classical research cycle, the researcher develops an idea with a solid theoretical foundation, elaborates the conceptual model and hypotheses, collects and analyses the required data and then 'writes it up' as a lucid and compelling paper or thesis. The paper then influences its readers and so contributes to the advancement of the science. In practice, such a linear process would be unlikely to be successful, especially if success is measured in terms of the influence the research achieves. Socially and politically relevant research requires engagement and iteration. Moreover, such a strictly sequential process would be isolating, particularly for an inexperienced researcher. 'Writing up' can be a lonely and painful experience if it is all left to the end, and many a PhD has foundered at this stage. In this chapter we consider the process and the key elements of successful research communication.

We begin with the assumption that the research aims to be policy relevant. Indeed, there is increasing pressure on researchers and research organizations to demonstrate their worth with evidence of impact. In academic work, evidence may be sought in the form of publication records, with increasing attention to citation indices and journal impact factors. There is also increasing emphasis on measuring research achievement in terms of uptake (evidence that influential people/organizations have used research information), influence (evidence of changes to policy or practice) and impact (evidence of actual livelihoods and/or environmental benefits 'on the ground') (Meagher et al, 2008). Contemporary understanding of the role of research in policy-making recognizes multiple and iterative pathways. Peer-reviewed publications remain a fundamental part of research communications, but there are ways of doing research and of reporting it that can make the research more effective. Moreover, there are other ways of communicating research that can complement and supplement journal articles to help get the message across.

The first step is to ensure that the research focus is relevant and important. It is a recurrent message throughout this book, but it is truly the key to success. When it comes to reporting, the basic guidance is simple: be clear about who you are talking to and what message you want to deliver, then tailor the writing accordingly. There is plenty of good advice regarding how to write. There are books, online resources and university writing centres (Box 14.2). But it is not always easy to know who the audience really is. Who are you trying to influence? What argument will move them? How can the research results and recommendations be presented most effectively? The chapter begins with a brief discussion of how research can influence policy, and specifically regarding the policy environment for international development research. We then discuss the research itself and the kinds of information and analyses that can be influential. Then we look at the writing process, with hints and tips on how to present empirical data in an accessible and informative way. Finally, we consider the publishing process, and again offer some suggestions to increase the effectiveness and the success of submissions.

Influencing policy

'*Policy* aims for continuity or change of a practice, including plans and their evolution when put into practice (that is, the "how" as well as the "what" of decisions)' (Shankland, 2000, cited in Crewe and Young, 2002, p3). Whether and how research influences policy is a well-established topic of inquiry for social scientists. In the 1950s, Lasswell suggested a model of the policy-making process as a series of stages during which information is rationally considered by policy-makers, but that model has been contested for at least 30 years (Crewe and Young, 2002). Modern concepts see it as a complex interplay between political interests, competing discourses and the agency of multiple actors (Keeley and Scoones, 1999; Crewe and Young, 2002). Research can inform the process at different stages. Research can also be ignored or misused at different stages! There may be very different worldviews among researchers and policy-makers, a cultural gap that prevents adequate use of research (Neilson, 2001). And of course, research is more likely to be used if it is politically expedient.

A source of frustration among researchers is that policy-makers selectively pick research results and arguments that fit their agenda.

Notwithstanding this attention to the impact of research on policy, there is still a certain amount of naïvety among some researchers who seem to assume that their responsibility starts and ends with publishing in the peer-reviewed press. The idea that published research will somehow 'inform policy' still seems to hold sway and the old linear model of policy formation is implicit in the way some researchers work.

How does research get to policy-makers? The Millennium Ecosystem Assessment (2003, p213) defined a policy-maker as, 'a person with power to influence or determine policies and practices at an international, national, regional, or local level'. David Kaimowitz, former director general of the Center for International Forestry Research (CIFOR), argued that researchers should know not only who their work was aimed at in a general sense, but also their names and email addresses (personal communication). Depending on the issue, it may be appropriate to focus at the level of officials who are deciding policy within government or other organizations, or it may be more effective to focus on lower levels where regulations are made or enforced or projects are implemented.

This book deals primarily with research regarding livelihoods and natural resources in developing countries and there are particular characteristics to keep in mind when communicating such research. Government decision-makers in poor countries often lack well-developed institutions to generate and use research and (partly for that reason) much of the research-based information they do have comes from other providers, such as the World Bank (Weiss, 2009). More to the point, government decision-makers may have little real influence on what happens on the ground. Government agencies in developing countries often lack the personnel and other resources to reach rural populations. Other kinds of organizations, such as private resource extraction and processing companies, the BINGOs (big international non-governmental organizations), bilateral donor agencies and other conservation and development organizations often have a larger presence and greater influence than government, especially in rural and remote areas. Research-based knowledge that influences the policies of such organizations can directly impact people's livelihoods and natural resources.

Policy-relevant research

You have already given considerable thought to the policy implications of your research while defining your research problem (see Chapter 3). Well-focused research will delve into a relevant and important issue and offer information and analyses to help to understand and address the problem.

Research may focus directly on the impact of policy. For example, research can predict (*ex ante*) or measure (*ex post*) the impact and distributional effects of policies or policy tools. Research can also provide knowledge for forming, implementing or contesting policy by: identifying and explaining trends; raising awareness of a problem; improving understanding of underlying causes of economic behaviour and environmental outcomes; contradicting conventional wisdom; identifying best practices; developing/influencing methods; or developing a useful theory or conceptual framework or model. By providing a clear theoretical explanation for a phenomenon or a conceptual framework for thinking about a problem, research can facilitate practical interventions.

Reaching the audience

It is not enough to publish a paper and hope for uptake. A citation analysis by Meho (2007, p32) noted: 'It is a sobering fact that some 90% of papers that have been published in academic journals are never cited. Indeed, as many as 50% of papers are never read by anyone other than their authors, referees and journal editors.' This suggests two important strategies for a researcher who wants to make a difference. First, when publishing in peer-reviewed journals, make sure the article is well targeted and well presented, so it will be read and used. Secondly, consider using other forms of communication to supplement and complement the journal articles – get the message across in different ways, to different sectors of the intended audience. Exposing the research more broadly can stimulate extra interest in the journal articles and increase the potential for impact.

Research impact studies find that the way the research is done strongly influences its impact (Carden, 2009). Engaging the intended audience during the research process helps to ensure the relevance of the research question, and it also prepares the audience to be more interested and receptive to the results and recommendations. Lomas (2000, p140) observed:

> *Researchers need to appreciate that decision making is not so much an event as it is a diffuse, haphazard, and somewhat volatile process. Similarly, decision makers need to recognize that research, too, is more a process than a product. Better links between research and decision making depends on the two communities finding points of exchange at more than the 'product' stage of each of their processes.*

O'Neil (2005, p762) identifies two key characteristics of influential research: intent and engagement. She notes: 'Research influence will only survive if research is designed from the start and carried out and translated to the policy people with a resolute and explicit and specific intent.' In addition: 'Where researchers form personal relationships with people in policy-making, their influence on policy is both more immediate and more lasting. Where those relationships fail to develop, influence is precarious or non-existent' (O'Neil, 2005, p762). This analysis does not negate the importance of focus and quality in the research, but it highlights the importance of engagement as another necessary element to help influence policy.

Spilsbury and Kaimowitz (2000) asked forest policy experts to identify publications that had been influential in national and international policy. They found little evidence that publications directly influence policy, but some were important in enhancing awareness and shaping conventional wisdom and policy narratives. The authors concluded that it was probably not the documents per se that had the impact, but rather the processes accompanying their creation. Research that targets or associates itself with major policy processes or powerful organizations has a better chance of having an impact and being recognized. They noted that being right is not necessary to have an impact – work that is later criticized or discredited by scientific peers can be highly influential in raising issues, shifting scientific debate and shaping policy outcomes, as in the example of a paper by Peters et al (1989) on non-timber forest products. But credibility is important, and this is at least partly determined by the reputations and track records of the authors, the prestige of the publishers and the influence of the organizations that sponsored the research and/or promoted the findings. They noted that research that tells policy-makers and opinion leaders what they want to hear has a better chance of being influential than work that goes against the current popular understanding. The overall recommendation was that policy researchers can increase their impact not only by providing good answers to the right questions but also by supplying these messages to the right people at the right time and in an appropriate format. The most influential researchers and institutions will be those who effectively build 'coalitions' to support their viewpoints in the policy arena and succeed in associating their work with well-funded initiatives.

People in the communities where the research has been done are still too often overlooked as an important audience for research outputs. Research-based knowledge, and the research processes themselves (especially participatory processes), can inform and empower stakeholders so that they can have more influence over policy and practice that affects them. Stakeholder participation helps ensure the relevance of questions and the appropriateness of answers and the way those answers are delivered and it helps inform and mobilize public opinion, even if only at the village scale (see also Chapter 2).

These are lessons that can be applied even in a PhD research project. It is well worth the effort to meet officials and staff in relevant government departments and project personnel working in the area. In this way, you can learn about their objectives and activities and about baseline information that may be available. You can test the questions and methods with people who know the situation and who may be able to help in adjusting and focusing the work so that it will be relevant and useful. By sharing research progress and ideas as they take shape, one can prepare the ground for when the full results are ready. A key audience for the research will already be aware of the work and interested to know the results, and the results are likely to be more easily accepted because the research design has incorporated ideas from the intended clients.

Media reports and other popular summaries can give research outputs wider reach. Many more readers will see a reference to a scientific report in a newspaper or magazine than will ever read the article. The message will be shaped by the way the journalist portrays it. Still, it is worth investing effort to get popular messages out, especially as a way to influence and to provide information and analysis to be used by civil society organizations. A combination of media and non-governmental organization (NGO) interest can create attention for a research report. Local newspaper articles, newsletter pieces, email updates, meeting or conference reports, presentations to interested stakeholders, and informal communications can all be valuable methods of reaching your audience.

There is also an increasing number of digest services that provide synopses of scientific articles.[1]

All of these kinds of communication help to convey your message and create awareness and interest in the research. All are part of the process of engagement. Not incidentally, they are all also part of the larger writing process. Each time you are forced to formulate and present ideas to an audience, in person or in print, you refines the message and develops the text that can be utilized later. It also creates opportunities for feedback that will help focus and increase the relevance of the message.

The scientific report

The peer-reviewed literature is important for communication within the scientific community and as a way to validate the scientific value of the research. The peer-review system helps assure the quality of scientific publications and, as discussed later, peer review is invaluable as a source of critique, new ideas and inspiration for authors. A journal article can influence the way other scientists think and how they do their own research. And, as Park (2009) notes, peer-reviewed publications are deeply embedded in the academic reward system. Almost all universities use publication records as indicators of productivity, and citation records are increasingly used as indicators of scientific worth.

There is a standard organizational structure for scientific writing that has stood the test of time for good reasons. Whether it is a PhD thesis or a journal article, the classic organization provides a logical and compelling structure for reporting research. The various sections may be compressed or expanded to meet the needs of the particular document. A PhD thesis, for example, needs a comprehensive and elaborated literature review section because part of the purpose of the document is to demonstrate the student's mastery of the field. A letter in the journal *Nature* must be far more concise; references are cited without any detail provided. Some journals want a separate background and rationale; others prefer to have that covered in the introduction. Some combine results and discussion while others separate them. The author needs to be familiar with the specific journal requirements and tailor the document accordingly, but the basics of building and presenting a good argument remain the same.

The report must be based on a clear, well-articulated research question. It should provide a strong rationale for investigating the research question and the policy relevance should be established early. The reader needs to understand and be convinced by the overall line of reasoning. References should be used to provide background information and evidence that will support the general and specific aspects of the argument. There should be a clear explanation regarding how data were gathered and how they were analysed. It is helpful to present results as simple informative graphics. Any assumptions and models need to be explained well. The goal is a clear, concise and well-argued paper with a self-contained summary, conclusions and explicit policy recommendations. Let us look at the elements in order:

Title

The title should be concise, catchy and informative. You want to attract the reader's interest and create a positive impression. It should indicate the subject and scope of the paper. Including keywords, especially words that are topical in the field, will help attract attention and search engine hits. It is not always possible to meet all of these criteria. A common ploy is to use puns or references to popular song or movie titles, or to give a twist to a popular saying. A title can be both catchy and informative by using a subtitle to explain or provide context for a more imaginative main title.

Here is a sample titles taken from CIFOR's POLEX list:

- 'From Mao to markets in China's forests'
- 'Globalizing local communities'
- 'The conservation of donors'
- 'Disturb forests for their own good'
- 'Certifying the little guys'
- 'Chainsaws in the drugstore'
- 'Will the eucalypts eat your children?'
- 'Filipinos "think locally, act locally"'

Abstract (or summary)

Aside from the title, the abstract is the first and possibly the only part of your paper that will be read. Considering the quick and often cursory way that editors screen manuscript submissions, the quality of the abstract may even determine if other readers will have the opportunity to read your paper.

It is still common to receive articles for review that have descriptive abstracts. Some even get published. A descriptive abstract outlines the topics covered in the article in a kind of elaborated table of contents, presumably so that the reader can decide whether or not to read the article. Such an abstract provides little information and suggests a lack of focus.

In comparison, an informative abstract provides details about the content of the article or report. A good abstract provides a condensed version of the main argument (usually 100–500 words), with an indication of the relevance and importance of the problem and the results, a clear problem statement, a brief description of the method and scope of the research and a summary of the results and the conclusions. The abstract alone should provide readers with a basic understanding of the research and its conclusions even if they choose not to read the article. It may also convince readers to read the rest of the article.

Introduction

The introduction should catch the reader's interest and attention and set the stage for the rest of the report. Starting with a question, a quote, a startling fact or a contradiction immediately gets the reader engaged and interested. Journalists often use a particular perspective to introduce an issue, and that can also be effective in a scientific article.

Try to introduce the main idea in the first sentence or paragraph. This gives the reader a sense of the direction of the paper so they can more easily absorb the background and context that follows. Sometime authors provide a great deal of background and build up to the main point, but readers (and editors) are impatient.

Background and rationale/literature review

Depending on the topic and the journal, the 'background and rationale' can be presented in a separate section or it can be provided in a condensed form in the introduction. This section draws on the literature to establish the relevance of the paper, to set the context and to provide support for the approach to be used. If another author has described the conditions or the policy environment or the culture in the study area, you can refer to that description to provide background and support for your own description of the context. If the research will engage in an ongoing debate, the paper should provide relevant references to introduce and delimit that debate and provide a foundation or a counterpoint for the current discussion. Every argument builds to some extent on previous work. By citing influential papers, you can efficiently convey well-supported ideas, while giving appropriate credit to the authors of those ideas and guiding the reader where to look for background. You can summarize the essence of the previous work without needing to present it in detail. Citations also help provide credibility for your work. By referring to other trustworthy sources, you reinforce the notion that the issue is important and that your work has a solid foundation. References are especially important to help frame a debate and the varying viewpoints in a discourse. By documenting the different perspectives and arguments you help the reader to understand your own arguments and conclusions. This is especially important when forming an opposing argument or attempting to disprove someone else's conclusions.

It is not helpful to have a long list of references if many of the references in the list are not germane to the argument. The references used should provide relevant and useful descriptions of the context, of the theoretical or empirical foundation for the argument in the paper or a clear counterpoint. Avoid using references that do not make a genuine contribution. (See also the discussion in Chapter 3 on literature reviews in research proposals – most of the points made are also valid for scientific papers.)

The problem statement

The introduction provides the basis for the problem statement. The problem statement may be included within the introduction (common) or in a separate section; regardless, it is a fundamental and necessary part of any research report.

Weak or absent problem statements are a frequent cause for rejection of article submissions.

The problem statement encapsulates the context for the research and serves as the basis for the questions the study aims to answer. The problem statement often consists of two parts: a knowledge-related problem originating in theory and a specific case of the problem manifested a real situation. As discussed in Chapter 3, the researcher will invest considerable effort in defining the research idea (problem) as part of the research design. A clear and concise problem statement presents the focus and the boundaries of the research. It should identify the variables being investigated and the relationship between the variables being investigated and the target population, and it should be presented in its simplest form.

> *In a sense, research is like dealing with a set of propositions in a debate or an argument adhering to the principles of logic. The purpose is to persuade or gain acceptance of the conclusion. To do so, it is essential for others to accept the first and all subsequent propositions. The problem statement is the first proposition, and we need to accept it before considering the next proposition.* (Hernon and Schwartz, 2007, p309)

Purpose and objectives

The purpose statement identifies how the research will contribute to solving the problem identified in the problem statement. It is a broad statement of desired outcomes or the general intentions of the research. It emphasizes what is to be accomplished, but not how it will be accomplished. The purpose statement is an aspirational goal that the research will contribute to, but that is outside the control of the research project.

The objectives define specifically what the research will achieve in a way that is realistic and measureable. It is common to indicate a general, overarching objective and a set of specific objectives. Each specific objective is an individual statement of intention, operationalized in the research matrix through specification of research questions, hypotheses, data needs and data analysis methods (Chapter 3). Reviewers and readers alike will expect these objectives to be satisfied by the research and explained in the article. Like the problem statement, purpose and objectives may be part of the introduction.

Methods

With the research question and specific objectives established, you should then provide a description of how the data were collected and analysed. This is the 'methods' section of the paper. Many manuscripts are submitted with weak methods sections. The best test of a description of methods is whether the explanation is sufficient for another researcher 'skilled in the art' to be able to replicate the study. The study site, study population, data collection and analysis methods should be described. The ease of access to online references makes it possible to provide thorough supporting information, but the description in the article should be sufficient. This way the reader can appreciate and understand the results and this builds trust. If particular methods were applied for each of the specific objectives, they should be presented in order.

Results

The results are the new evidence that the researchers use to support their argument. The results section should present a summary of the relevant data and analyses, and here again it is important to tailor the presentation to the intended audience. Good analysis is often handicapped in its presentation because the authors have not taken the vital last step of thinking about what the audience cares about and expressing the results in those terms (Koomey, 2001). Some audiences (and some journals) will expect and appreciate having regression outputs presented in tables with coefficients, standard errors and marginal effects. Other audiences may find that level of detail distracting or difficult to understand. Graphical presentations that illustrate and highlight the key patterns or trends may be more accessible and effective.

There are three basic ways to present results: text, tables and figures. If there are few results to report, it is not necessary to use a table or a figure – simply report the results in the text. For example, to report average household forest dependency for two villages it may be tempting to provide a table such as Table 14.1.

The same information could be conveyed more efficiently in a sentence without breaking the flow of the narrative. However, if there are several variables to report for more than one of the units of analysis (villages), then a table or a figure can be very useful indeed (for a comprehensive discussion see Day and

Table 14.1 *Average forest income dependency*[2]

Village	*Forest dependency*
Village A	20%
Village B	30%

Box 14.1 *Resist the temptation of the pie chart*

The pie chart (or circle graph) is commonly used to illustrate proportions. In technical terms, the arc length of each sector (and consequently its central angle and area), is proportional to the quantity it represents. But people are not good at judging angles: we underestimate acute angles (angles less than 90°) and overestimate obtuse angles (angles greater than 90°) (Robbins, 2005). They also use a lot of ink so they are expensive to print. To illustrate the weakness of pie charts, we present Figure 14.1, plotted from a fictional data set we put together for illustration purposes.

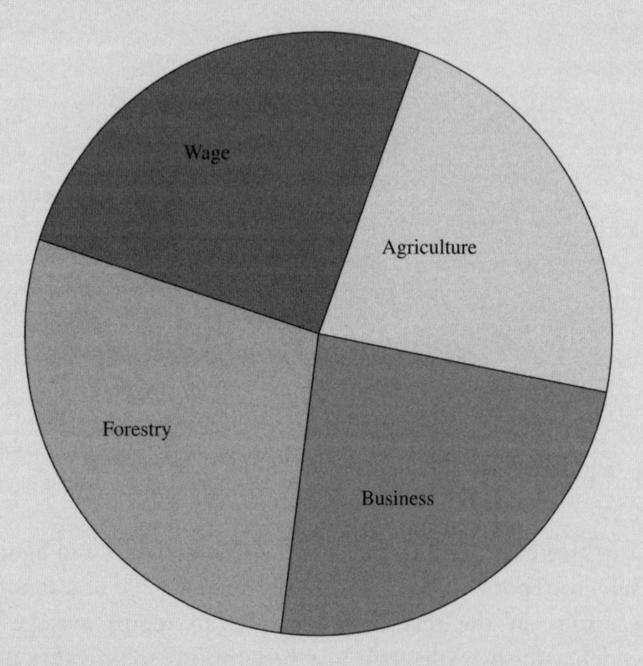

Figure 14.1 *Average share of total household income: Pie chart*
Source: Illustration data from authors

In looking at a chart such as this, one may want to know what categories have the largest and smallest shares, and also some sort of ranking. Given time, the reader would figure this out. However, if this figure were part of a presentation where the presenter spent the typical 1–2 minutes on a slide, there is a good chance the audience would have some difficulties quickly getting answers to those interesting questions. One could argue that the slices could be sorted, or labelled with percentages, or the most important sector could be 'exploded' from the main chart. Any of these actions would help the reader interpret the chart, but they will also add clutter, and clutter never aids understanding.

Contrast the pie chart with the two charts below:

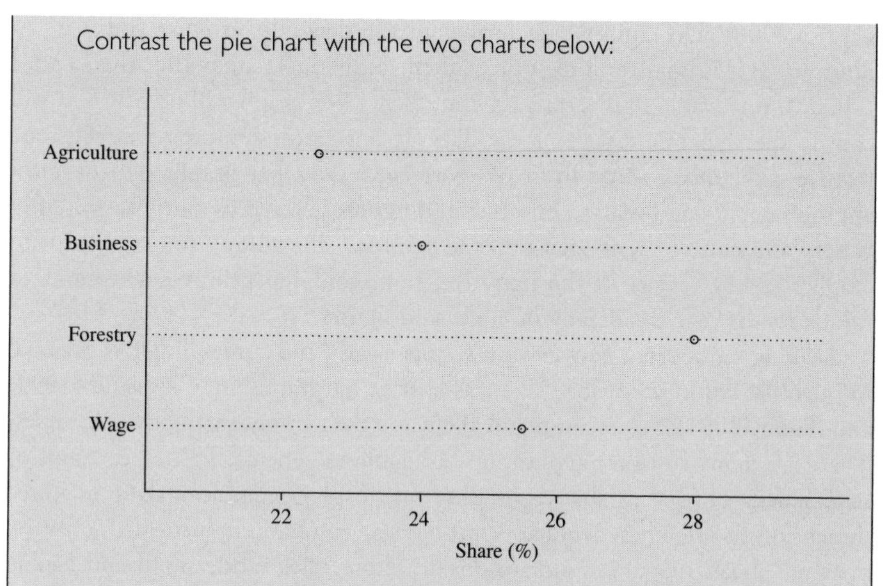

Figure 14.2 *Average share of total household income: Dot plot chart*
Source: Illustration data from authors

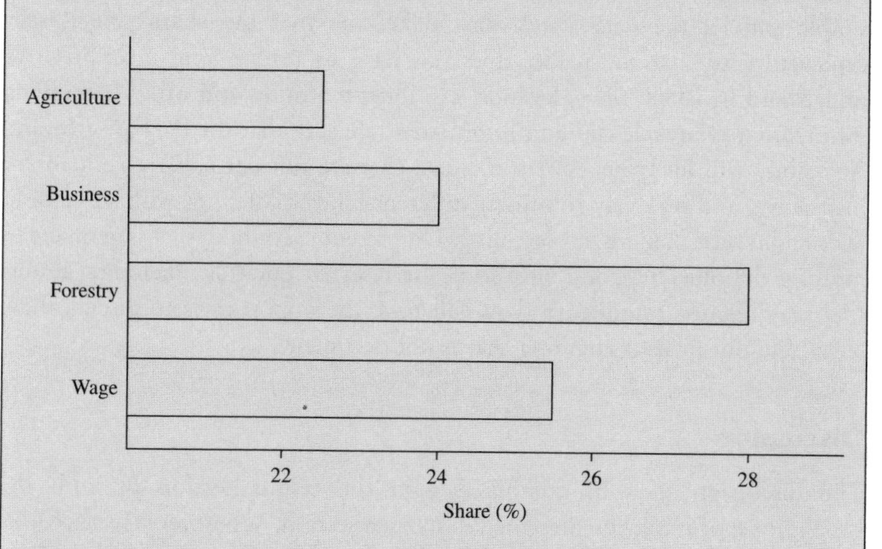

Figure 14.3 *Average share of total household income: Bar chart*
Source: Illustration data from authors

It is likely that the audience will more immediately understand both charts. It is also likely that the bars do not need to be labelled for the audience to decode the approximate number of shares. Both graphs are easier to read because they provide information positioned on a common scale. For more on people's ability to interpret graphs, see Robbins (2005).

Gastel, 2006). Deciding which format to use depends on what you want to illustrate. If it is important to appreciate the individual data points, then a table is best. If one wants to illustrate a trend, then a line graph will be best as it will convey the dynamic aspect of the data. If it is most important to illustrate relative differences, some form of chart (such as a bar graph) will be most appropriate. A combination of tables and figures selected by purpose will offer variety and complement each other to reinforce the study. You may want to emphasize key results in the narrative, but avoid repetitious descriptions of minor results that are already in tables and figures.

Making figures that convey your results clearly and compellingly is essential in scientific communication. The classic texts on graphs are Cleveland (1994) and Tufte (2001); we recommend them for their many instructive examples. There are many software applications available to generate figures, cartoons or schematics. However, the temptation to display graphical data in three dimensions is often best avoided – just because you have sophisticated software, does not mean that three dimensions are more easily understood and legible than two dimensions.

Presenting results is about communication, and clarity always wins. Keep it simple and let the data stand out. Make sure that the chart/table is self-explanatory so that the reader does not have to read a long description to understand it. Make titles, legends, captions, footnotes and other identifying information clear, relevant and informative. Be sure that the data are relevant. Any study will likely generate some data that are not particularly germane or instructive, and one may run many different kinds of analysis with some dead ends and results that are not revealing. Do present everything that is germane to meeting the objectives and answering the research question, including results that may seem contradictory (you will have the opportunity to discuss these later). Do not present anything that is not pertinent.

Discussion

The discussion may be combined with the results section or with the conclusions, or it may be presented in its own section. Wherever it is placed, its purpose is to elaborate the argument and draw out the key messages. Sometimes inexperienced researchers get themselves trapped reiterating the results in terms of the empirical facts. The discussion needs to lift the argument to the next level with interpretation and exploration of what the results mean in the context of the research question and the overall research problem. It should also inform the reader of the study's scope and limitations – identifying issues that were not investigated and problems that were encountered – and indicate how far the findings can be generalized.

Conclusions and recommendations

This is where the argument is concluded. It should be a concise and clearly focused section that draws on the discussion to present answers to the questions that were asked. If those conclusions indicate recommendations for changes (or continuation) of policy or practice, this is the place to make those recommendations. Be sure that whatever conclusions and recommendations are offered are sound and well-supported. As Samet (1999, p435) advised:

> *[Do not] write a weak last paragraph. This is where authors often lose control, offering sometimes naïve policy recommendations or generic calls for more research (possibly in support of their next grant). Manuscripts need an ending, but must go out with restraint.*

Acknowledgements

It is good practice to acknowledge individuals, groups and organizations that contributed to the research in the form of ideas, information, advice, or technical or financial support. Be sure to specifically acknowledge each person or organization (they will probably look). However, recognize that not everyone will want to be mentioned. They may disagree with something or for some other reason they might prefer not to be associated with the report, so you should seek permission. Acknowledging financial support gives due recognition to the funders and it provides credibility by association for the research.

The writing process

As already emphasized, scientific writing, like science itself, is an iterative, incremental process. In some ways, the researcher starts writing the paper the day the idea is born (Bourne, 2005). The research proposal identifies and articulates a research goal and a policy angle. The details will be tested, re-evaluated and refined as the researcher gains knowledge and experience of the situation. Keep asking questions in policy-relevant terms. This thinking, and the writing that goes with it, all contribute to the final products.

It is normally very helpful to start the writing with an annotated outline early in the process and then build on it incrementally as the research and analysis progresses. Include key arguments, results, interpretations and references. This outline can be a valuable tool in the writing process. It is also handy to share and build an annotated outline with co-authors, or with your supervisor, as a way to ensure that everyone is 'on the same page'.

Another helpful step is to draft sections as you go. Some are easy and natural, such as descriptions of the background and rationale, methods and site. It is never too early to begin notes and draft sections of the discussion, as ideas arise.

As discussed above, engaging with stakeholders, and publishing interim reports and synopses will help your research to have impact. Aim for different outputs: local newspaper article; newsletters of relevant NGOs; email updates to key clients of research; synopses presented in terms that are locally relevant. All of this helps to anticipate the final outputs. It also provides a solid foundation of writing that can be drawn upon for the more formal academic outputs.

Getting published and getting through

The work is not over when you complete your manuscript draft. Next comes the submission, and you need to pay careful attention to the key elements that will get your article past the first hurdle. The editor of a premier international development journal highlighted this when he reported that he personally reviews every submission to decide whether or not to send it for review. He spends three to five minutes making that decision and he rejects 90 per cent of submissions at that stage. Like most readers, he focuses on the abstract, introduction and conclusion to judge whether the subject is relevant and sufficiently novel, whether the argument is cogent and well-supported, and whether the presentation is interesting. Unlike readers, the editor will also see your cover letter, which should be informative but not too long. The cover letter gives you the opportunity to communicate the relevance and importance of your research in a way that reinforces the abstract and introduction. Editors are also keenly aware of page length limitations. If your article can make the same point in less space, it will have a better chance of being published.

If the editor thinks the article has potential, it will be sent to peer reviewers who will assess its relevance and quality. The feedback can be devastating; reviewers are not always kind or constructive, but all feedback is useful. Good reviewers may provide critiques of the underlying assumptions, the problem definition, the methods, arguments, organization or any other aspect of the paper. They may provide helpful suggestions to improve the paper and they may suggest alternate ways of analysing the data or other references that would inform the argument. This feedback is a great gift and all comments and suggestions should be carefully considered. Some reviewers may make comments that seem to misunderstand the paper completely and they might be rude or mean. These reviews can be useful as well, so try to avoid defensiveness and consider why the reviewer responded that way. Presumably, the editor selected smart and experienced scholars as reviewers. If they misunderstood something in the paper,

Box 14.2 *Writing resources*

Writing

- 'Writing for change', www.idrc.ca/IMAGES/books/WFC_English/WFC_English/sitemap.html. This excellent resource provides opportunities to learn how to write effectively, as well as how to write for scientific publications and for advocacy.
- Strunk, W. (1918) *Elements of Style*, private printing, Ithaca, NY, www.bartleby.com/141. A classic overview of the style rules that facilitate clear communication.

University websites regarding academic writing: These university websites all provide a wealth of information on a wide range of writing topics.

- OWL at Purdue: http://owl.english.purdue.edu/owl.
- University of Wisconsin-Madison Writing Centre Writer's Handbook: http://writing.wisc.edu/Handbook/index.html.
- University of Toronto Writing Centre: 'Advice on academic writing': www.writing.utoronto.ca/advice.

Editing

- 'How to prevent and fix problems in papers' section in 'How to get published': www.law.upenn.edu/cpp/alumni/jobseekers/GetPublishedSpecialReportACADEMICWORD.pdf. Provides advice on how to address common publishing issues, including recommendations for how to choose an editor.
- 'Effective editing' section of 'Writing for change': www.idrc.ca/IMAGES/books/WFC_English/WFC_English/effedi1.html. Provides a step-by-step approach to editing, including exercises to develop editing skills.

Journalistic writing for scientists

- 'Communicating science: Tools for scientists and engineers': http://communicatingscience.aaas.org/Pages/newmain.aspx. Addresses four areas of communication: 'Communication basics', 'Working with reporters', 'Public outreach' and 'Multimedia', plus provides links to other resources.

Presentations

- 'Oral presentation advice': http://pages.cs.wisc.edu/~markhill/conference-talk.html. Provides advice and a generic outline for an oral conference presentation.

- 'Creating an effective conference presentation': www.kon.org/karlin.html. Provides general advice on preparing and presenting conference posters and papers.
- 'Chronology of your successful conference presentation': www.thinkoutsidetheslide.com/Conference_Success.pdf. Gives a step-by-step timeline to planning a conference presentation.
- Page, M. (2003) *The Craft of Scientific Presentations: Critical Steps to Succeed and Critical Errors to Avoid*, Springer, New York. Provides a framework for successful presentations.

Poster presentations

- 'Poster presentations': http://library.buffalo.edu/asl/guides/bio/posters.html. Provides tips on designing effective posters, as well as numerous links to more information on the topic.

then other readers might also misunderstand. If they found the paper irritating, other readers might also be irritated. Reconsider the paper from that perspective and assess whether an idea could be communicated more clearly or if the presentation could be more convincing or engaging.

You can now revise and improve the paper based on feedback from the editor and the reviewers. If the paper was rejected outright, you can reformat it for another journal as part of the revision process. If, however, the paper was accepted 'with revisions', you will need to respond to all of the major comments. You do not have to agree with them all but you must demonstrate why your approach is correct. When you resubmit, provide a cover letter that itemizes reviewers' suggestions and indicates how you have responded to them. Some journals send resubmitted papers back to the original reviewers; others send to fresh reviewers. In either case, an itemized list of main comments and responses demonstrates your seriousness and professionalism and makes the subsequent review easier.

Conclusions

Research communication is an iterative process. Engaging with stakeholders early in the process helps to focus and refine the research question(s) to ensure relevance. It also helps to create and prepare an audience for the research results and recommendations. The audience for international conservation-oriented and development-oriented research is broad and messages may need to be targeted specifically for different sub-sectors. An individual research project may yield practical recommendations that are useful at the local level, policy guidance for local or national government, feedback or advice that will support project implementation, and/or confirmation or contradiction of international policy or programme directions. A solid argument based on novel results and tailored for the appropriate audience can have a powerful influence. Time-tested presentation styles and organizational structures provide a good foundation for research publishing but other kinds of communications will both complement the traditional outputs and facilitate their preparation. By making writing and other research communications an integral part of the research process, it is possible to increase the effectiveness of the research and the efficiency of the writing.

Key messages

- Successful research communication must be targeted to address the interests and needs of the intended audience.
- The writing process is iterative and integral to the research process itself – start early and build your messages as you work.
- A good presentation is about developing and defending an argument, with a clear concise illustration of your data and analyses.

Notes

1 Some relevant digests for research on international conservation and development issues include Eldis (www.eldis.org), Mongabay (www.mongabay.com), Community Forestry e-news (www.recoftc.org/site/) and CIFOR's POLEX listserv (www.cifor.cgiar.org/Knowledge/Polex/).
2 The table also lacks the obvious and yet often omitted piece of information: the sample size. Now, it is possible that this would be included in the text but this is a problem. A good table should have all the information needed for a reader to understand it.

References

Bourne, P. (2005) 'Ten simple rules for getting published', *PloS Computational Biology*, vol 1, no 5, www.ploscompbiol.org/article/info:doi/10.1371/journal.pcbi.0010057, last accessed 5 February 2011

Carden, F. (2009) *Knowledge to Policy: Making the Most of Development Research*, International Development Research Centre, Ottawa, Canada

Cleveland, W. S. (1994) *The Elements of Graphing Data*, Wadsworth Advanced Books and Software, Monterey, CA

Crewe, E. and Young, J. (2002) 'Bridging research and policy: Context, evidence and links', ODI Working Papers 173, September, www2.dwaf.gov.za/webapp/

ResourceCentre/Documents/Research_And_Development/wp173%5B1%5D.pdf, last accessed 5 February 2011

Day, R. A. and Gastel, B. (2006) *How to Write and Publish a Scientific Paper*, Greenwood Press, Santa Barbara, CA

Hernon, P. and Schwartz, C. (2007) 'What is a problem statement?', *Library and Information Science Research*, vol 29, pp307–309, available at www.lis-editors.org/bm~doc/editorial-problem-statement.pdf, last accessed 5 February 2011

Keeley, J. and Scoones, I. (1999) 'Understanding environmental policy processes: A review', IDS Working Paper no 89, Institute of Development Studies, Brighton

Koomey, J. G. (2001) *Turning Numbers in Knowledge: Mastering the Art of Problem Solving*, Analytics Press, Oakland, CA

Lomas, J. (2000) 'Connecting research and policy', http://portals.wi.wur.nl/files/docs/ppme/lomas_e.pdf, last accessed 5 February 2011

Meager, L., Lyall, C. and Nutley, S. (2008) 'Flows of knowledge, expertise and influence: A method for assessing policy and practice impacts from social science research', *Research Evaluation*, vol 17, no 3, pp163–173, doi:10.3152/095820208X331720

Meho, L. I. (2007) 'The rise and rise of citation analysis', *Physics World*, vol 20, no 1, pp32–36

Millennium Ecosystem Assessment (2003) 'Glossary', in *Ecosystems and Human Well-being: A Framework for Assessment*, www.maweb.org/documents/document.59.aspx.pdf, p213, last accessed 5 February 2011

Neilson, S. (2001) *IDRC in the Public Policy Process: A Strategic Evaluation of the Influence of Research on Public Policy*, www.idrc.ca/evaluation/policy, last accessed 5 February 2011

O'Neil, M. (2005) 'What determines the influence that research has on policy-making?', *Journal of International Development*, vol 17, no 6, pp761–764

Park, J. (2009) 'Motivations for web-based scholarly publishing: Do scientists recognize open availability as an advantage?', *Journal of Scholarly Publishing*, vol 40, no 4, pp343–369, doi:10.3138/jsp.40.4.343

Peters, C. M., Gentry, A. H. and Mendelsohn, R. O. (1989) 'Valuation of an Amazonian rainforest', *Nature*, vol 339, no 6227, pp655–656

Robbins, N. B. (2005) *Creating More Effective Graphs*, John Wiley & Sons, Hoboken, NJ

Samet, J. M. (1999) 'Dear author: Advice from a retiring editor', *American Journal of Epidemiology*, vol 150, no 5, pp433–436

Shankland, A. (2000) 'Analysing policy for sustainable livelihoods', *IDS Research Report*, vol 49, pp1–49

Spilsbury, M. J. and Kaimowitz, D. (2000) 'The influence of research and publications on conventional wisdom and policies affecting forests', *Unasylva*, vol 51, no 4, pp3–10

Tufte, E. R. (2001) *The Visual Display of Quantitative Information*, Graphic Press LLC, Cheshire

Weiss, C. H. (2009) 'Foreword', in Carden, F. (ed) *Knowledge to Policy: Making the Most of Development Research*, International Development Research Centre, Ottawa, Canada

Index